Reaction-Diffusion Equations

Reaction-Diffusion Equations

The Proceedings of a Symposium Year on Reaction-Diffusion Equations organized by the Department of Mathematics, Heriot-Watt University, 1987–1988

Edited by

K. J. BROWN and A. A. LACEY
Department of Mathematics, Heriot-Watt University

CLARENDON PRESS · OXFORD 1990

Oxford University Press, Walton Street, Oxford OX2 6DP
Oxford New York Toronto
Delhi Bombay Calcutta Madras Karachi
Petaling Jaya Singapore Hong Kong Tokyo
Nairobi Dar es Salaam Cape Town
Melbourne Auckland
and associated companies in
Berlin Ibadan

Oxford is a trade mark of Oxford University Press

Published in the United States
by Oxford University Press, New York

© Oxford University Press 1990

All rights reserved. No part of this publication may be reproduced,
stored in a retrieval system, or transmitted, in any form or by any means,
electronic, mechanical, photocopying, recording, or otherwise, without
the prior permission of Oxford University Press

This book is sold subject to the condition that it shall not, by way
of trade or otherwise, be lent, re-sold, hired out or otherwise circulated
without the publisher's prior consent in any form of binding or cover
other than that in which it is published and without a similar condition
including this condition being imposed on the subsequent purchaser

British Library Cataloguing in Publication Data
Brown, K. J.
Reaction-diffusions equations.
1. Differential equations
I. Title II. Lacey, A. A. (Andrew Alfred), 1953–
515. 35
ISBN 0–19–853378–0

Library of Congress Cataloging in Publication Data
(Data available)

Printed in Great Britain
by Biddles Ltd.
Guildford and King's Lynn

Contents

Preface   vii

Addresses of Contributors   ix

J. Bebernes and *Alberto Bressan*   1
Blowup for some reactive-Euler induction models

G. C. Wake, J. B. Burnell, J. G. Graham-Eagle and *B . F. Gray*   25
A new scaling of a problem in combustion theory

Masayasu Mimura   39
Patterns, waves and interfaces in excitable reaction-diffusion systems

J. Smoller and *A. Wasserman*   59
Reduced equivariant Conley index and applications

Nicholas D. Alikakos and *William R. McKinney*   75
Remarks on the equilibrium theory for the Cahn-Hilliard equation
in one space dimension

Jonathan Bell   95
Excitability behaviour of myelinated axon models

Chris Cosner   117
Eigenvalue problems with indefinite weights and reaction-diffusion models
in population dynamics

Paul C. Fife and *Xiao Geng*   139
Mathematical aspects of electrophoresis

R. A. Gardner   173
Topological methods for the study of travelling wave solutions
of reaction-diffusion systems

Jesús Hernández   199
Maximum principles and decoupling for positive solutions
of reaction-diffusion systems

Preface

Nonlinear reaction-diffusion equations arise in models for a large number of physical, chemical, biological and other problems. The solutions also display a wide variety of behaviour including the formation of patterns, non-existence through blow-up, and development of travelling waves. As a consequence of the diverse forms of equations and range of possible behaviour, techniques for gaining understanding of the problems must also cover a wide spectrum.

This book contains a collection of articles which survey recent developments over the whole area of reaction-diffusion equations. The contributions indicate both the wide range of situations in which reaction diffusion equations can arise, for example, biology (nerve propagation), electrochemistry, combustion (ignition) and ecology as well as the wide range of mathematical techniques which are being brought to bear on such problems, e.g., classical partial differential equation techniques such as comparison principles, nonlinear functional analysis and topological index theory.

The individual authors were participants in a reaction-diffusion symposium year held in 1987-88 at Heriot-Watt University, Edinburgh, and are well-known for their work in the field. The year was a section of a large programme generously supported by the Science and Enginnering Research Council, towards whom our gratitude is extended, as part of the initiative on non-linear systems.

The editors would also like to thank all those who took part in the symposium year, particularly the contributors to this book of proceedings.

A A Lacey   K J Brown

Edinburgh, May 1990

Addresses of Contributors

Nicholas D. Alikakos
Mathematics Department
University of Tennessee
Knoxville, TN 37996
USA

J. Bebernes
Department of Mathematics
University of Colorado
Boulder, CO 80309
USA

Jonathan Bell
Department of Mathematics
University at Buffalo, SUNY
Buffalo, New York 14214
USA

Alberto Bressan
Department of Mathematics
University of Colorado
Boulder, CO 80309
USA

J. B. Burnell
Applied Mathematics Division
Department of Scientific
and Industrial Research
Wellington
New Zealand

Chris Cosner
Department of Mathematics
and Computer Science
University of Miami
Coral Gables,
Florida 33124,
USA

Paul C. Fife
Department of mathematics
University of Utah
Salt Lake City
Utah 84112
USA

J. G. Graham-Eagle
Department of Theoretical
and Applied Mechanics
University of Auckland
New Zealand

Xiao Geng
Brown University
Providence
Rhode Island 02912
USA

R. A. Gardner
Department of Mathematics
and Statistics
University of Massachusetts
Amherst
Massachusetts MA 01003
USA

B. F. Gray
School of Chemistry
Macquarie University
Sydney
Australia

Jesús Hernández
Departamento de Matemáticas
Universidad Autónoma
28049 Madrid
Spain

William R. McKinney
Department of Mathematics
North Carolina State University
Raleigh, NC 27965
USA

Masayasu Mimura
Department of Mathematics
Hiroshima University
Hiroshima
Japan

J. Smoller
Department of Mathematics
The University of Michigan
Ann Arbor, Michigan 48109
USA

G. C. Wake
Department of Mathematics
and Statistics
Massey University
Palmerston North
New Zealand

A. Wasserman
Department of Mathematics
The University of Michigan
Ann Arbor, Michigan 48109
USA

Blowup for some reactive-Euler induction models

J. Bebernes
Alberto Bressan

University of Colorado

1 Introduction

The evolution of thermal explosions in gaseous systems depends on the interaction between chemical heat release, conductive thermal losses, and the effects of compressibility. The latter factor can accelerate reaction rates in constant volume systems where compression heating plays a role (Bebernes and Bressan 1982; Kassoy and Poland 1983; Bebernes and Kassoy 1988). In unconfined systems however, the conversion of some thermal energy to kinetic energy may retard the appearance of thermal runaway. Systems in which conductive losses are unimportant will inevitably explode, perhaps faster than diffusive systems. In this sense it is important to be able to predict which physical processes control the evolution of an exothermic reaction in a specific gaseous system. In this paper, we present the results of our recent studies which provide a rational basis for deciding the correct induction model for the given physical system and analyze these models mathematically(Kassoy et al. 1989a, b; Bebernes and Kassoy 1989; Bressan, in press;).

Consider a reactive, viscous, heat-conducting, compressible gas in an equilibrium state defined by the dimensional quantities $p_0 = p(0, x)$, $\rho_0 = \rho(0, x)$, $T_0 = T(0, x)$, $y_0 = y(0, x)$, and $u_0 = u(0, x)$ which represent pressure, density, temperature, concentration, and velocity, respectively.

At time $t = 0$, assume a small initial disturbance is created on a length scale L. Define $\bar{x} = x/L$ as the new position vector. Let $\bar{t} = t/t_R$ be the new time scale where t_R is a reference time to be determined later. Nondimensionalize the system variables, letting $\bar{p} = p/p_0$, $\bar{\rho} = \rho/\rho_0$, $\bar{T} = T/T_0$, $\bar{y} = y/y_0$, and $\bar{u} = u/(L/t_R)$. Assume a single one-step irreversible reaction which has a rate law described by Arrhenius kinetics. The complete combustion system can then be written in nondimensional form, where the bar notation has been dropped,

$$(1.1) \begin{cases} \rho_t + \nabla \cdot (\rho u) = 0 \\ \rho(u_t + u \cdot \nabla u) = -\frac{1}{\gamma}(\frac{t_R}{t_A})^2 \nabla p + \Pr(\frac{t_R}{t_c})\mu[\Delta u + \frac{1}{3}\nabla(\nabla \cdot u)] \\ \rho c_v (T_t + u \cdot \nabla T) = \gamma(\frac{t_R}{t_A}) \nabla \cdot (k \nabla T) - (\gamma-1)p(\nabla \cdot u) \\ \qquad\qquad +2\mu[\frac{\gamma(\gamma-1)\Pr t_A^2}{t_R t_c}][\mathcal{D}:\nabla \otimes u - \frac{1}{3}(\nabla \cdot u)^2] \\ \qquad\qquad + t_R B \rho y \exp[-\frac{1}{\varepsilon T}] \\ \rho(y_t + u \cdot \nabla y) = \text{Le}(\frac{t_R}{t_c}) \nabla \cdot (\rho D \nabla y) - t_R B \rho y \exp(-\frac{1}{\varepsilon T}) \\ p = \rho T \end{cases}$$

where $\bar{\mu} = \mu/\mu_0$, $\bar{D} = D/D_0$, $\bar{c}_p = c_p/c_{p_0}$, $\bar{c}_v = c_v/c_{v_0}$, $\bar{k} = k/k_0$, and $\bar{K} = K/K_0$ where $K = k/\rho c_p$ is the thermal diffusivity, c_p and c_v are the specific heats. Also, $\gamma = c_{p_0}/c_{v_0}$ is the gas parameter, $\varepsilon = RT_0/E$ is the nondimensional inverse of the activation energy, $\Pr = c_{p_0}\mu_0/k_0$, the Prandtl number, $\text{Le} = D_0/K_0$ the Lewis number, $c_0 = (\gamma RT_0)^{1/2}$ the initial sound speed, $t_A = L/c_0$ the acoustic time scale, $t_c = L^2/K_0$ the conduction time scale and $\bar{h} = hy_0/c_{v_0}T_0$ is the nondimensional heat of reaction.

2 Induction period models

As in Jackson *et al.* (1987), assume that $\Pr = O(1)$, $\text{Le} = O(1)$, $h = O(1)$ and that $\varepsilon \ll 1$. Using the method of activation energy asymptotics, we seek simpler models of the combustion process. In (1.1c,d), the reaction terms contain an expression of the form $\exp(-\frac{1}{\varepsilon T})$. For $\varepsilon \ll 1$, an induction period theory can be described in terms of the perturbed variables

$$(2.1) \begin{cases} \rho = 1 + \varepsilon m, \quad p = 1 + \varepsilon P, \quad T = 1 + \varepsilon \theta \\ u = \varepsilon v, \quad y = 1 - \varepsilon c, \end{cases}$$

where we assume that the initial temperature disturbance is $O(\varepsilon)$. If $O(\varepsilon)$ terms are ignored, we obtain the induction model for a gaseous system from (1.1) using

(2.1):

$$(2.2)\begin{cases} m_t + \nabla \cdot v = 0 \\ v_t = -\frac{1}{\gamma}(\frac{t_R}{t_A})^2 \nabla P + \Pr(\frac{t_R}{t_c})\mu[\Delta v + \frac{1}{3}\nabla(\nabla \cdot v)] \\ \theta_t = t_R B h \varepsilon^{-1} e^{-1/\varepsilon} e^{\theta} + \gamma(\frac{t_R}{t_A})\Delta\theta - (\gamma - 1)\nabla \cdot v \\ \qquad + 2\gamma(\gamma-1)\frac{t_A^2}{t_R \cdot t_c}\varepsilon \Pr\mu[-\frac{1}{3}(\nabla \cdot v)^2 + \{\nabla \otimes v + (\nabla \otimes v)^T\} : \nabla \otimes v] \\ c_t = t_R B \varepsilon^{-1} e^{-1/\varepsilon} e^{\theta} + \text{Le}(\frac{t_R}{t_c})\Delta c \\ P = m + \theta. \end{cases}$$

The induction model (2.2) contains three time scales t_R, t_A, and t_c which depend on the particular thermochemical system with the reference time t_R yet to be specified. The character of the induction models depends intimately on the ratios formed from these three time scales. We will consider initial temperature disturbances on a macroscopic length scale so that $t_A/t_c \ll 1$. If we assume that the perturbation temperature θ and the concentration c variations are caused by the chemical reaction process, then for ε small there should be a balance of the accumulation terms θ_t and c_t in (2.2) with the reaction terms involving e^{θ}. It follows that the reference time can be defined by

$$(2.3) \qquad t_R = \frac{\varepsilon e^{1/\varepsilon}}{B}$$

which represents the chemical time for a reaction initiated at T_0 multiplied by ε. The three time scales are now completely defined and the reduced induction models depends on their ratios.

The first case to be considered is that for

$$(2.4) \qquad \frac{t_R}{t_c} \equiv a = O(1).$$

then the induction momentum, energy, and species equations of (2.2) can be written as

$$(2.5) \begin{cases} (\frac{t_A}{t_c})^2 v_t = -\frac{a^2}{\gamma}\nabla P + a(\frac{t_A}{t_c})^2 \Pr\mu[\Delta v + \frac{1}{3}\nabla(\nabla \cdot v)] \\ \theta_t = he^{\theta} + a\gamma\Delta\theta - (\gamma - 1)\nabla \cdot v \\ c_t = e^{\theta} + \text{Le}\,a\Delta c \end{cases}$$

Since we are assuming that the initial disturbances are spatially macroscopic so that $t_A/t_c \ll 1$, we have from the inductive momentum equation (2.5a) that $P = P(t)$ to a first approximation. Combining (2.2e), the mass equation (2.2a), and the energy equation (2.5b),

(2.6) $$\theta_t = \frac{h}{\gamma}e^\theta + a\Delta\theta + \frac{\gamma-1}{\gamma}P'(t).$$

For a bounded container Ω since the total mass must be conserved, $\int_\Omega \rho(t,x)dx = \text{vol}(\Omega)$ which implies $\int_\Omega m(t,x)dx = 0$ and hence

$$P(t) = \frac{1}{\text{vol}\Omega}\int_\Omega \theta(t,x)dx.$$

We can thus rewrite (2.6) as

(2.7) $$\theta_t - a\Delta\theta = \delta e^\theta + \frac{\gamma-1}{\gamma}\frac{1}{\text{vol}\Omega}\int_\Omega \theta_t(t,x)dx$$

and impose initial-boundary conditions of the type

(2.8) $$\begin{cases} \theta(0,x) = \theta_0(x), & x \in \Omega \\ \theta(t,x) = 0, & (t,x) \in \partial\Omega \times (0,\infty). \end{cases}$$

This model (2.7)-(2.8) with the last term representing the effects of spatially homogeneous gas compression was originally derived by Kassoy and Poland(1983) and was analyzed by Bebernes and Bressan(1982).

If the ratio $t_R/t_c \equiv a \ll O(1)$ so that the reaction time is much shorter than the conduction time, then (2.2b,c,d) can be written

(2.9) $$\begin{cases} v_t = -\frac{1}{\gamma}\frac{(t_R/t_c)^2}{(t_A/t_c)^2}\nabla P + \text{Pr}a\mu[\Delta v + \frac{1}{3}\nabla(\nabla\cdot v)] \\ \theta_t = he^\theta - (\gamma-1)\nabla\cdot v + a\gamma\Delta\theta + 2\gamma(\gamma-1)\text{Pr}\varepsilon\mu\frac{(t_A/t_c)^2}{a}. \\ \qquad [-\frac{1}{3}(\nabla\cdot v)^2 + \{\nabla\otimes v + (\nabla\otimes v)^T\} : \nabla\otimes v] \\ c_t = e^\theta + \text{Le}\cdot a\cdot \Delta c. \end{cases}$$

Because $a = o(1)$, viscous, conductive, and diffusive effects are weak. Three subcases are of interest, all of which lead to *reactive-Euler explosions*.

I) For $t_A \ll t_R \ll t_c$, then from (2.9a) $P = P(t)$ to a first approximation and the energy equation becomes

(2.10) $$\begin{aligned} \theta_t &= \frac{h}{\gamma}e^\theta + \frac{\gamma-1}{\gamma}P'(t) \\ &= \frac{h}{\gamma}e^\theta + \frac{\gamma-1}{\gamma}\frac{1}{\text{vol}\Omega}\int_\Omega \theta_t(t,x)dx \end{aligned}$$

where Ω is a bounded container.

II) For $O(t_R) = t_A \ll t_c$, to first order the momentum equation (2.9a) becomes

$$\text{(2.11)} \qquad v_t = -\frac{1}{\gamma} \frac{a^2}{(t_A/t_c)^2} \nabla P$$

and (2.2) reduces to

$$\text{(2.12)} \qquad \begin{aligned} \theta_t - \frac{\gamma-1}{\gamma} P_t &= \frac{h}{\gamma} e^\theta \\ v_t + \frac{1}{\gamma} \frac{a^2}{(t_A/t_c)^2} \nabla P &= 0 \\ \nabla \cdot v + \frac{1}{\gamma} P_t &= \frac{h}{\gamma} e^\theta \end{aligned}$$

III) For $t_R \ll t_A \ll t_c$, (2.9a) reduces to $v_t = 0$ or $v = v(x)$. This implies that inertial confinement of the heated gas is dominant. Aspects of short time inertial confinement have been discussed by Clarke and Cant(1985), Dold(1988), Jackson and Kapila(1985), and Jackson et al. (1989).

3 The first reactive-Euler model

For an arbitrary bounded container $\Omega \subset \mathbf{R}^N$, the reactive-Euler model (2.10) can be written as

$$\text{(3.1)} \qquad \phi_t = \delta e^\phi + \frac{\gamma-1}{\gamma} \frac{1}{\text{vol}\Omega} \int_\Omega \phi_t(t,x) dx$$

with

$$\text{(3.2)} \qquad \phi(0,x) = \phi_0(x)$$

assuming $\phi_0(x)$ is continuous and bounded on Ω. By integrating (3.1) over Ω, we see that (3.1) is equivalent to

$$\text{(3.3)} \qquad \phi_t = \delta e^\phi + \beta \int_\Omega e^\phi dx$$

where $\beta = \frac{(\gamma-1)\delta}{\text{vol}\Omega}$.

The IBVP (3.3)-(3.2) has a unique nonextendable solution $\phi(t,x)$ on $\Omega \times [0,\sigma)$ where $\sigma = +\infty$ or $\sigma < \infty$ with $\lim_{t \to \sigma^-} \sup\{\phi(x,t) : x \in \Omega\} = \infty$.

The initial value problem

$$\text{(3.4)} \qquad a' = \delta e^a, \quad (t,x) \in \Omega \times (T,0)$$

$$\text{(3.5)} \qquad a(0,x) = \phi_0(x), \quad x \in \bar{\Omega}$$

has the explicit solution

$$\text{(3.6)} \qquad a(t,x) = -\ln[e^{-\phi_0(x)} - \delta t]$$

which blows up in finite time $T = \delta^{-1}\exp(-\phi_0(x_m))$ where x_m is any point in Ω at which $\phi_0(x)$ attains its absolute maximum. Since $a(t,x)$ is a lower solution for (3.3)-(3.2), the solution $\phi(t,x)$ satisfies

$$\phi(x,t) \geq -\ln[e^{-\phi_0(x)} - \delta t]$$

and hence $\phi(t,x)$ blows up in finite time σ with $\sigma \leq T$.

To get more information about $\phi(t,x)$, consider the implicit representation

(3.7) $$\phi(t,x) = a(\tau(t),x) + B(\tau(t))$$

where $a(\tau,x)$ is the solution of (3.4)-(3.5) and $\tau(t), B(\tau)$ are scalar functions to be determined. As given in (3.7), $\phi(t,x)$ is a solution of (3.3)-(3.2) if and only if

(3.8) $$\tau' = e^{B(\tau)}, \quad \tau(0) = 0$$

(3.9) $$B' = \beta\int_\Omega e^{a(x,\tau)}dx = \beta\int_\Omega [e^{-\phi_0(x)} - \delta\tau]^{-1}dx, \quad B(0) = 0.$$

By intergrating (3.9), (3.8) can be solved by quadrature to get

(3.10) $$B(\tau) = \frac{\beta}{\delta}\int_\Omega [a(\tau,x) - \phi_0(x))]dx = \frac{\beta}{\delta}\int_\Omega \ln[\frac{e^{-\phi_0(x)}}{e^{-\phi_0(x)} - \delta\tau}]dx$$

and τ satisfies

(3.11) $$\tau' = \exp[\frac{\beta}{\delta}\int_\Omega \ln[\frac{e^{-\phi_0(x)}}{e^{-\phi_0(x)} - \delta\tau}]dx], \quad \tau(0) = 0$$

which can be solved by quadrature.

From (3.7), we thus have

Theorem 3.1 *The number σ is the blow-up time for the solution $\phi(t,x)$ of (3.1)-(3.2) if and only if $\tau(\sigma) = T$ is the blow-up time for the solution $a(\tau,x)$ of (3.4)-(3.5), and thus $\sigma = \tau^{-1}(\frac{1}{\delta}e^{-\phi_0(x_m)})$ where x_m is any point in Ω at which ϕ_0 has an absolute maximum.*

By considering (3.7) and (3.10), we can observe that $\phi(t,x)$ only blows up at these points x_m at which $\phi_0(x)$ has its absolute maximum provided that $B(\tau(\sigma)) < \infty$. This is true if and only if $\int_\Omega a(\tau(\sigma),x)dx < \infty$ which in turn is true provided that $\int_\Omega \ln[e^{-\phi_0(x)} - e^{-\phi_0(x_m)}]dx > -\infty$. Similarly, $\phi(t,x)$ blows up everywhere in Ω at σ if and only if $B(\tau(\sigma)) = \infty$. Thus,

Theorem 3.2 (a) *The solution $\phi(t,x)$ of (3.1)-(3.2) blows up only at these points x_m of Ω at which $\phi_0(x)$ has its absolute maximum if and only if*

(3.12) $$\int_\Omega \ln[e^{-\phi_0(x)} - e^{-\phi_0(x_m)}]dx > -\infty.$$

(b) *The solution $\phi(t,x)$ blows up everywhere in Ω at σ if and only if*

(3.13) $$\int_\Omega \ln[e^{-\phi_0(x)} - e^{-\phi_0(x_m)}]dx = -\infty.$$

The integral in (3.12) is finite if there is at most a finite number of critical points $x_m \in \Omega$ at which ϕ_0 has an absolute maximum and if at each x_m $\phi_0(x)$ is strictly concave down and analytic in a neighbourhood of x_m. In this case, blow-up occurs only at those x_m at which ϕ_0 has an absolute maximum. If on the other hand ϕ_0 is too flat in a neighbourhood of an x_m, then blow-up occurs everywhere in Ω.

A second method for representing the solution $\phi(t,x)$ of (3.1)-(3.2) is to set

$$(3.14) \qquad \Phi(t,x) = \phi(t,x) - \frac{\gamma-1}{\gamma} \frac{1}{\text{vol}\Omega} \int_\Omega \phi(t,x) dx.$$

Then Φ satisfies

$$(3.15) \qquad \Phi_t = \delta F_t e^\Phi$$

with

$$(3.16) \qquad \Phi(0,x) = \phi_0(x) - \frac{\gamma-1}{\gamma} \frac{1}{\text{vol}\Omega} \int_\Omega \phi_0(x) dx$$

where

$$(3.17) \qquad F_t = \exp[\frac{\gamma-1}{\gamma} \frac{1}{\text{vol}\Omega} \int_\Omega \phi(t,y) dy], \quad F(0) = 0.$$

By integrating (3.15) and using (3.17), we find that $\phi(t,x)$ can be expressed as

$$(3.18) \qquad \phi(t,x) = \frac{\gamma-1}{\text{vol}\Omega} \int_\Omega \ln\frac{1}{G} dy + \ln\frac{1}{G}$$

where $G(t,x) = ke^{-\phi_0(x)} - \delta F(t)$, $k = \exp[\frac{\gamma-1}{\gamma} \frac{1}{\text{vol}\Omega} \int \phi_0(y) dy]$. Note then that the blow-up time σ for ϕ is given from (3.18) by $F(\sigma) = \frac{ke^{-\phi_0(x_m)}}{\delta}$.

Since $P_0(t) = \frac{1}{\text{vol}\Omega} \int_\Omega (\phi(t,x) - \phi_0(x)) dx$, we have from (3.18)

$$(3.19) \qquad P_0(t) = \frac{\gamma}{\text{vol}\Omega} \int_\Omega \ln\frac{1}{G(t,y)} dy - \frac{1}{\text{vol}\Omega} \int_\Omega \phi_0(y) dy.$$

From (3.18) and (3.19), we have

$$(3.20) \qquad \phi(t,x) = \phi_0(x) + \frac{\gamma-1}{\gamma} P_0(t) - \ln(1 - \frac{\delta}{k} e^{\phi_0(x)} F(t))$$

from which we can conclude that the temperature evolves from the initial value $\phi_0(x)$ through a purely time dependent term related to the homogeneous pressure increase and a logarithmic evolution term with spatial dependence which has a shape-preserving property.

4 The second reactive-Euler model

In one spatial dimension, the reactive-Euler model (2.12) can be written as

$$\vartheta_t - \frac{\gamma-1}{\gamma} P_t = \delta e^\vartheta$$
(4.1)
$$v_t + \frac{1}{\gamma}\left(\frac{a'}{t_A/t_c}\right)^2 P_x = 0$$
$$v_x + \frac{1}{\gamma} P_t = \delta e^\vartheta.$$

Setting $a = \frac{\gamma-1}{\gamma}$, $b = \delta$, $c = \frac{1}{\gamma}\left(\frac{a'}{t_A/t_c}\right)^2$, $d = \frac{1}{\gamma}$, then (4.1) becomes a semilinear hyperbolic system of the form

(4.2)
$$\begin{cases} \vartheta_t - a P_t = b e^\vartheta \\ v_t + c P_x = 0 \\ v_x + d P_t = b e^\vartheta \end{cases}$$

which is a model for the behaviour of a reactive gas where viscous and diffusive effects are sufficiently weak. Here the variables P, v and ϑ denote the perturbations from an equilibrium state of pressure, velocity and temperature respectively, while a, b, c, d are positive constants. We first study the system (4.2) in the bounded domain $I = [-1, 1]$, with initial and boundary conditions

(4.3) $\qquad P(0, x) = \bar{P}(x), \quad v(0, x) = \bar{v}(x), \quad \vartheta(0, x) = \bar{\vartheta}(x),$

(4.4) $\qquad\qquad\qquad v(t, 1) = v(t, -1) = 0.$

Here $\bar{P}$, $\bar{v}$, $\bar{\vartheta}$ are continuous functions on I, with $\bar{v}(1) = \bar{v}(-1) = 0$. The presence in (4.2) of a reaction term which grows exponentially with the temperature allows solutions to become unbounded within finite time. Our main concern is the location of the blow-up and the description of the profile of a solution as the explosion time is approached. In §6 we prove that a generic solution of (4.2)-(4.4) blows up at a single point. In §7, assuming that the initial conditions are smooth and that initial temperature has a sufficiently large and "well focused" maximum (in a sense to be specified later), we determine the asymptotic profile of a solution close to the blow-up point. A topological argument (Hartman 1982, p278) combined with standard comparison techniques shows that, in a suitable set of rescaled coordinates, the solution has a nontrivial nonsingular limit. This asymptotic limit depends on finitely many parameters, determined by the initial conditions. For a single parabolic equation, a similar rescaling of coordinates was studied by Giga and Kohn (1985) and later by Bebernes et al. (1987) for an exponential nonlinearity. We now supply a rigorous proof to the formal computations on the reactive-Euler model which appeared in (Jackson et al. 1987; Dold 1988).

5 Preliminaries for the second reactive-Euler model

The change of variables

(5.1)
$$\begin{cases} z_1 = \vartheta - aP \\ z_2 = \frac{a}{2}[P + (cd)^{-1/2}v] \\ z_3 = \frac{a}{2}[P - (cd)^{-1/2}v] \end{cases}$$

tranmorms (4.2), (4.4) into the system

(5.2)
$$\begin{cases} z_{1,t} = Ae^{z_1+z_2+z_3} \\ z_{2,t} + \lambda z_{2,x} = Be^{z_1+z_2+z_3} \\ z_{3,t} - \lambda z_{3,x} = Be^{z_1+z_2+z_3} \end{cases}$$

with boundary conditions

(5.3) $\qquad z_2(t,1) = z_3(t,1), \quad z_2(t,-1) = z_3(t,-1).$

Here $A = b$, $B = ab/2d$, $\lambda = (c/d)^{1/2}$. Call $E \subset \mathcal{C}(I, \mathbf{R}^3)$ the Banach space of all continuous functions $z = (z_1, z_2, z_3)$ from $[-1, 1]$ into $\mathbf{R}^3$ which satisfy

$$z_2(1) = z_3(1), \quad z_2(-1) = z_3(-1),$$

with the usual norm

$$\|z\| = \max\{|z_i(x)|;\ |x| \le 1, i = 1, 2, 3\}.$$

For a given set of initial conditions

(5.4) $\qquad\qquad z_i(0, x) = \bar{z}_i(x),$

with $\bar{z} \in E$, it is well known (Courant and Hilbert 1962; Martin 1976) that the system (5.2), (5.3) has a unique forward solution z defined on some maximal interval $[0, T)$. Using the semigroup notation, we write $S^t(\bar{z}) = z(t, \cdot) \in E$ for the value of this solution at time t. Observe that either $T = +\infty$ or $\|S^t(\bar{z})\| \to \infty$ as $t \to T^-$. In this second case, we call $T = T(\bar{z})$ the blow-up time for the initial conditions (5.4). A point $\bar{x} \in [-1, 1]$ is a blow-up point if there exist sequences $x_n \to \bar{x}$, $t_n \to T^-$ such that $|z(t_n, x_n)| \to +\infty$. The set of blow-up points, given the initial data $z(0, \cdot) = \bar{z}(x)$, is denoted by $B\ell(\bar{z})$. On the space E, define the ordering

$$u \preceq v \quad \text{iff} \quad u_i(x) \le v_i(x) \quad \forall x \in I,\ i = 1, 2, 3.$$

By a standard comparison theorem, if $u \preceq v$, then $S^t(u) \preceq S^t(v)$ for all $t \geq 0$ at which both are defined. In particular, $u \preceq v$ implies $T(u) \geq T(v)$.

In a metric space Y, $B(y, \delta)$ denotes the open ball centred at y with radius δ. We recall that a set-valued map F is upper semicontinuous if, for all $\bar{z}$ and $\varepsilon > 0$, there exists $\delta > 0$ such that

$$z \in B(\bar{z}, \delta) \Rightarrow F(z) \subseteq B(F(\bar{z}), \varepsilon).$$

Here $B(F(\bar{z}), \varepsilon)$ denotes the neighbourhood of radius ε around the set $F(\bar{z})$. If Y is a complete metric space, a subset $R \subseteq Y$ is residual in Y (i.e., of second category) if and only if its complement $Y \setminus R$ lies in the union of countably many closed nowhere dense sets. A property $\mathcal{P}$ of elements of Y is generic if it holds true for all y in a residual subset of Y. In the next section we prove that the property "$B\ell(\bar{z})$ is a singleton" is generic in the space E. This is a mathematically precise way of expressing the fact that for "most" initial conditions $\bar{z} \in E$ the corresponding solution of (5.2)-(5.3) blows up at a single point. All of our results will be stated for the system (5.2), which is more tractable than (4.2). Performing the change of variables inverse of (5.1), analogous results for the original system (4.2) can be easily obtained.

6 Properties of the blow-up set

As a preliminary, observe that for every initial condition $\bar{z} \in E$ the solution z of (5.2),(5.3) blows up in finite time. Indeed, define

(6.1) $$c^+ = \max\{A, B\}, \quad c^- = \min\{A, B\},$$

(6.2) $$m^+ = \max\{\bar{z}_i(x);\ x \in I, i = 1, 2, 3\},$$
$$m^- = \min\{\bar{z}_i(x);\ x \in I, i = 1, 2, 3\}.$$

Comparing the components of z with the solutions of the scalar Cauchy problems

(6.3) $$y' = c^{\pm} e^{3y}, \quad y(0) = m^{\pm}$$

we deduce

(6.4) $$-\tfrac{1}{3}\ln[e^{-3m^-} - 3c^- t] \leq z_i(t, x) \leq -\tfrac{1}{3}\ln[e^{-3m^+} - 3c^+ t]$$

for $t > 0$, $x \in I$, $i = 1, 2, 3$. Hence the solution z blows up at some time $T = T(\bar{z})$ with

(6.5) $$(3c^+ e^{3m^+})^{-1} \leq T(\bar{z}) \leq (3c^- e^{3m^-})^{-1}.$$

Observe, however, that all solutions can be extended backwards for all times $t \in (-\infty, 0]$. Indeed, another simple comparison argument yields

$$m^- + c^+ e^{3m^+} t \leq z_i(t, x) \leq m^+ \quad \forall t < 0.$$

In this section we study the dependence of the blow-up time and of the blow-up set of the initial data and prove that, generically, solutions blow up at a single point.

Proposition 6.1 *The map $\bar{z} \to T(\bar{z})$ is continuous on E.*

Proof. Fix $\bar{z} \in E$. If $\tau < T(\bar{z})$, then $\|S^t(\bar{z})\| < \infty$ and $S^{-\tau}$ provides a homeomorphism of a bounded neighbourhood V of $S^\tau(\bar{z})$ onto a neighbourhood U of $\bar{z}$. This implies $T(\bar{u}) > \tau$ for all $\bar{u} \in U$, hence

$$(6.6) \qquad \liminf_{\bar{u} \to \bar{z}} T(\bar{u}) \geq T(\bar{z}).$$

To prove the converse inequality, fix $\varepsilon > 0$ and call $z = z(t,x), v = v(t,x)$ the solutions of (5.2),(5.3) with initial conditions $\bar{z}, \bar{v}$ respectively. For all t,x we have

$$(6.7) \qquad v_i(t,x) \geq \min\{\bar{v}_i(x); \ |x| \leq 1, \ i = 1,2,3\}.$$

Define m^- as in (6.2) and choose $\tau < T(\bar{z})$ and K large enough so that

$$(6.8) \qquad e^{1-K} e^{2-2m^-} < \varepsilon c^-,$$

$$(6.9) \qquad \max\{z_i(\tau,x); \ |x| \leq 1, \ i = 1,2,3\} > K.$$

As before, $S^{-\tau}$ provides a homeomorphism between a neighbourhood V of $S^\tau(\bar{z})$ and a neighbourhood U of $\bar{z}$. By possibly shrinking U and V, it is not restrictive to assume that

$$(6.10) \qquad \|\bar{v} - \bar{z}\| < 1, \ \|S^\tau(\bar{v}) - S^\tau(\bar{z})\| < 1 \qquad \forall \bar{v} \in U.$$

From (6.9), (6.10) we deduce

$$(6.11) \qquad \max\{v_i(\tau,x); \ |x| \leq 1, \ i = 1,2,3\} > K - 1 \ \forall \bar{v} \in U.$$

Assume that this maximum is $v_j(\tau, \bar{x})$. Integrating the j-th equation in (5.2) along the j-th characteristic line through $(\tau, \bar{x})$ and estimating the other components v_i, $i \neq j$ by $v_i(t,x) \geq m^- - 1$ we obtain

$$(6.12) \qquad \frac{d}{dt} v_j(\tau + t, \bar{x} + \lambda_j t) \geq c^- e^{-2m^- - 2 + v_j}.$$

Here $\lambda_1 = 0$, $\lambda_2 = \lambda$, $\lambda_3 = -\lambda$. If $j = 2, 3$, the corresponding characteristic may hit the boundary of $[-1,1]$. In such cases, the estimate can be continued along the reflected characteristic, through the same boundary point. A comparison with the solution of the Cauchy problem

$$y' = c^- e^{2m^- - 2 + y}, \ y(0) = K - 1$$

yields the estimate

$$(6.13) \qquad v_j(\tau + t, \bar{x} + \lambda_j t) \geq -\ln[e^{1-K} - c^- e^{2m^- - 2} t].$$

By (6.13),(6.8), every solution v with initial data $\bar{v} \in U$ blows up within time $\tau + \varepsilon$. This implies

$$\limsup_{\bar{v} \to \bar{z}} T(\bar{v}) \leq T(\bar{z}),$$

completing the proof.

Proposition 6.2 *The map $\bar{z} \to B\ell(\bar{z})$ is upper semicontinuous, compact valued, from E into $[-1, 1]$.*

Proof. It is clear that set $B\ell(\bar{z})$ is closed, hence compact. The upper semicontinuity of the map $B\ell$ will be established by proving that for each $\bar{z} \in E$ and each $\bar{x} \notin B\ell(\bar{z})$ there exists $\varepsilon > 0$ and a neighbourhood U of $\bar{z}$ such that

(6.14) $\qquad [\bar{x} - \varepsilon, \bar{x} + \varepsilon] \cap B\ell(\bar{v}) = \phi \quad \forall \bar{v} \in U$.

Let $\bar{z}, \bar{x}$ be given. Since $\bar{x} \notin B\ell(\bar{z})$, there exists $\delta > 0$ and L such that the solution $z = z(t, x)$ of (5.2)-(5.4) satisfies

(6.15) $\qquad z_i(t, x) < L - 1 \quad \forall i = 1, 2, 3, \ |x - \bar{x}| \le \delta, \ 0 \le t < T(\bar{z})$.

Set $\varepsilon = \min\{e^{3L}/7c^+, \delta/3\lambda\}$, with c^+ as in (6.1), and define $\tau = T(\bar{z}) - \varepsilon$. Then $S^{-\tau}$ provides a homeomorphism from a neighbourhood V of $S^\tau(\bar{z})$ onto a neighbourhood U of $\bar{z}$. By possibly shrinking U and V, in view of (6.15) and of Proposition 6.1, we assume that

(6.16) $\qquad v_i(\tau, x) < L \quad \forall i = 1, 2, 3, \ |x - \bar{x}| \le \delta,$

for all solutions v with initial conditions $\bar{v} \in U$, and that

(6.17) $\qquad T(\bar{v}) < T(\bar{z}) + \varepsilon \quad \forall \bar{v} \in U$.

Comparing the values of v_i with the solution of the Cauchy problem

(6.18) $\qquad y' = c^+ e^{3y}, \quad y(\tau) = L$

from (6.16) it follows

(6.19) $\qquad v_i(t, x) \le -\frac{1}{3} \ln(3c^+(\tau - t) + e^{-3L})$

for $t \in [\tau, T(\bar{v})]$, $|x - \bar{x}| \le \delta - \lambda(t - \tau)$. Since $T(\bar{v}) < \tau + 2\varepsilon$, (6.19) and the choice of ε imply that, for every initial condition $\bar{v} \in U$, the corresponding blow-up set does not contain the interval $[\bar{x} - \varepsilon, \bar{x} + \varepsilon]$.

Theorem 6.1 *The set E^* of all $\bar{z} \in E$ such that $B\ell(\bar{z})$ consists of a single point is residual in E.*

Proof. For every integer $m \ge 1$ define the set

$$A^m = \{\bar{z} \in E, \ \text{diam} B\ell(\bar{z}) \ge m^{-1}\},$$

where $\text{diam } V = \sup\{|x - y|; \ x, y \in V\}$ denotes the diameter of the set V. Clearly $E^* = E \setminus \cup_{m \ge 1} A_m$. It thus suffices to show that each A_m is closed and nowhere dense.

To see that A_m is closed, assume $v_n \in A_m$ for all $n \ge 1$, $v_n \to u$. Let x_n, $y_n \in B\ell(v_n)$ with $|x_n - y_n| \ge m^{-1}$. Choosing a suitable subsequence, we can assume that $x_{n'} \to \bar{x}$, $y_{n'} \to \bar{y}$. By Proposition 6.2 , $\bar{x}, \bar{y} \in B\ell(u)$, hence diam $B\ell(\bar{u}) \geq |\bar{x} - \bar{y}| \geq m^{-1}$ and $u \in A_m$.

It remains to prove that the complement A_m^c of A_m is everywhere dense. Fix $u \in E$ and let $\bar{x} \in B\ell(u)$. Define $\varepsilon = 1/9m$, $\delta = \varepsilon/\lambda$. Let $\varphi : [-1,1] \to [0,1]$ be a continuous function such that $\varphi(x) = 1$ if $|x - \bar{x}| \leq 2\varepsilon$, $\varphi(x) = 0$ if $|x - \bar{x}| \geq 3\varepsilon$. For each $n \geq 1$ define the functions $v_n = (v_{n_1}, v_{n_2}, v_{n_3})$ and $\bar{w}_n$ by setting

$$v_{n_i}(x) = u_i(T(u) - \delta, x) + n^{-1}\varphi(x), \quad \bar{w}_n = S^{-T+\delta}(v_n).$$

Clearly $v_n \to S^{T-\delta}(u)$ and $\bar{w}_n \to u$ as $n \to \infty$. We claim that $\bar{w}_n \notin A_m$ for every $n \geq 1$. Call $w_n(t,x)$ the solution of (5.2), (5.3) with initial conditions $u_n(x)$. Then

$$w_n(T(u) - \delta, x) = v_n(x) > u(T(u) - \delta, x)$$

for each $n \geq 1$, $|x - \bar{x}| \leq 2\varepsilon$. By the continuity of u with respect to time, there exists $\delta' \in (0, \delta)$ such that

(6.20) $\qquad w_{n_i}(T(u) - \delta, x) \geq u_i(T(u) - \delta + \delta', x)$

for $|x - \bar{x}| \leq 2\varepsilon$. From (6.20) we deduce

(6.21) $\qquad w_{n_i}(t, x) \geq u_i(t + \delta', x)$

whenever $t \geq T(u) - \delta$, $|x - \bar{x}| \leq 2\varepsilon - \lambda(t - T(u) + \delta)$.

Since $\bar{x}$ is a blow-up point for u, (6.21) implies $T(u_n) \leq T(u) - \delta'$. All blow-up points of u_n however are contained inside the set $\{x; |x - \bar{x}| \leq 4\varepsilon\}$, because $\bar{w}_n(t, x) = u(t, x)$ whenever $|x - \bar{x}| > 3\varepsilon + \lambda(t - T(u) + \delta), T(u) - \delta \leq t < T(\bar{w}_n)$. Hence diam $B\ell(\bar{w}_n) < m^{-1}$ and $\bar{w}_n \notin A_m$, for all $n \geq 1$.

7 Asymptotic estimates

The aim of this section is to describe the asymptotic profile of a solution of (5.2) as the blow-up time is approached. Since what matters here is just the local behaviour, for simplicity we neglect the boundary conditions (5.3) and work with the Banach space $\mathcal{C}^3(\mathbf{R}, \mathbf{R}^3)$ of functions $z = (z_1, z_2, z_3)$ which are three times continuously differentiable on $\mathbf{R}$. Furthermore, we assume $A + 2B = 1$. This condition is clearly not restrictive, because it can always be achieved by the time rescaling $t' = (A + 2B)t$.

Theorem 7.1 *There exist a nonempty open set $U \subset \mathcal{C}^3(\mathbf{R}, \mathbf{R}^3)$ such that, if $\bar{z} \in U$, the corresponding solution z of (5.2) – (5.4) blows up at an isolated point x_0. Moreover, if $T = T(\bar{z})$ is the blow-up time, there exist constants $\Omega < 0$ and z_i^∞ such that $z_1^\infty + z_2^\infty + z_3^\infty = 0$, with the following property. For every $\varepsilon > 0$ there exists $\delta > 0$ such that*

(7.1)
$$|z_1(t,x) - z_1^\infty + A\ln[(T-t) - \tfrac{\Omega}{2}(x-x_0)^2]| < \varepsilon$$

$$|z_i(t,x) - z_i^\infty + B\ln[(T-t) - \tfrac{\Omega}{2}(x-x_0)^2]| < \varepsilon$$

($i = 2, 3$), whenever $T - \delta \leq t < T$, $|x - x_0| < \delta$.

Further estimates can be easily deduced from (7.1). For example, the temperature $\vartheta = z_1 + z_2 + z_3$ satisfies

$$\lim_{t \to T^-} [\vartheta(t, x) + \ln((T - t) - \frac{\Omega}{2}(x - x_0)^2)] = 0$$

uniformly on every domain of the form $|x| \leq \psi(T - t)$, ψ being any continuous function with $\psi(0) = 0$. Moreover, for $x \neq x_0$ the limit values of $z_i(t, x)$ as $t \to T^-$ are well defined. In particular, the final temperature profile $\vartheta(T, \cdot)$ satisfies

(7.2) $\quad \lim_{x \to x_0} [\vartheta(T, x) + 2 \ln |x - x_0| + \ln(-\Omega/2)] = 0$.

A more precise version of the above theorem will be actually proved, providing computable conditions on the initial data $\bar{z}$ which imply (7.1). As a preliminary, observe that, given any $\alpha, \beta, \gamma > 0$, one can always find $\tilde{\tau}$ large enough so that the following conditions hold.

C1) $e^{\tilde{\tau}/6} > \max\{2\lambda, 12\beta\lambda/\alpha, 4\gamma\lambda, e^{1/2}\}$

C2) The solution (μ, m, M, D) of the Cauchy problem

(7.3) $\quad \begin{cases} \dot{\mu} = -4me^{-\tau/6} - \mu/2 & \mu(\tilde{\tau}) = \gamma/2 \\ \dot{m} = 2me^{-\tau/3} & m(\tilde{\tau}) = -\beta/2 \\ \dot{M} = 8m^2 e^{-\tau/3} & M(\tilde{\tau}) = -2\alpha \\ \dot{D} = 2\beta + 8\beta^2 e^{-\tau/3} - D & D(\tilde{\tau}) = 2\beta \end{cases}$

satisfies

(7.4) $\quad \mu(\tau) < \gamma, \quad m(\tau) > -\beta, \quad M(\tau) < -\alpha, \quad D(\tau) < 3\beta$

for all $\tau \geq \tilde{\tau}$.

Theorem 7.2 Let $\alpha, \beta, \gamma, \tilde{\tau}$ be constants for which the conditions $C1), C2)$ hold. Let $\bar{z} = (\bar{z}_1, \bar{z}_2, \bar{z}_3)$ be a C^3 function such that $\bar{\vartheta} = \bar{z}_1 + \bar{z}_2 + \bar{z}_3$ attains a local maximum $\hat{\vartheta} > \tilde{\tau} + 1$ at a point $\hat{x}$. In addition, assume that

i) $\max |(\bar{z}_i)_x| < e^{(\hat{\vartheta}-1)/2}\gamma/2$

ii) $\max |(\bar{z}_i)_{xx}| < 2\beta e^{\hat{\vartheta}-1}$

iii) $-e^{\hat{\vartheta}-1}\beta/2 \ < \ \sum_i \min(\bar{z}_i)_{xx} \ < \ \sum_i \max(\bar{z}_i)_{xx} \ < \ -2\alpha e^{\hat{\vartheta}+1}$,
the max and min being all taken over the set $|x - \hat{x}| \leq 2e^{2(1-\hat{\vartheta})/3}$. Then the solution z of (5.2), (5.4) blows up at an isolated point x_0, at some time T, with

$$|\hat{x} - x_0| \leq e^{2(1-\hat{\vartheta})/3}, \quad e^{-1-\hat{\vartheta}} \leq T \leq e^{1-\hat{\vartheta}}.$$

Moreover, the estimates (4.1) hold with $-\beta \leq \Omega \leq -\alpha$.

We outline here the main arguments in the proof, while the details will be worked out in §§8, 9. Since all solutions of (5.2) have components which remain uniformly bounded from below, the blow-up occurs precisely when their sum $\vartheta = z_1 + z_2 + z_3$ becomes unbounded. If ϑ blows up at $t = T$, $x = x_0$, we expect that a nontrivial, nonsingular asymptotic limit can be obtained using the rescaled variables

(7.5) $$\tau = -\ln(T-t), \quad \eta = (x-x_0)/\sqrt{T-t},$$

(7.6) $$\begin{cases} u = z_1 + A\ln(T-t) \\ v = z_2 + B\ln(T-t) \\ w = z_3 + B\ln(T-t) \end{cases}$$

(7.7) $$S = u + v + w = \vartheta + \ln(T-t).$$

In these new variables, (5.2) takes the form

(7.8) $$\begin{cases} u_\tau + \tfrac{\eta}{2} u_\eta = A(e^S - 1) \\ v_\tau + (\tfrac{\eta}{2} + \lambda e^{-\tau/2}) v_\eta = B(e^S - 1) \\ w_\tau + (\tfrac{\eta}{2} - \lambda 3^{-\tau/2}) w_\eta = B(e^S - 1), \end{cases}$$

(7.9) $$S_\tau + \tfrac{\eta}{2} S_\eta + \lambda e^{-\tau/2}(v_\eta - w_\eta) = e^S - 1.$$

The new initial conditions, at $\tau_0 = -\ln T$ are

(7.10) $$u(\tau_0, \eta) = \bar{u}(\eta) = \bar{z}_1(x_0 + \eta T^{1/2}), \quad v(\tau_0, \eta) = \bar{v}(\eta) = \bar{z}_2(x_0 + \eta T^{1/2}),$$
$$w(\tau_0, \eta) = \bar{w}(\eta) = \bar{z}_3(x_0 + \eta T^{1/2}).$$

We remark, however, that the exact values of T and x_0 are not known *a priori*. Instead, one knows the point $\hat{x}$ where the maximum value $\hat{\vartheta}$ of $\vartheta = z_1 + z_2 + z_3$ is initially attained. Of course, we expect $x_0 \simeq \hat{x}$ and $T \simeq e^{-\hat{\vartheta}}$, but equality need not hold, in general. For this reason, we have to consider a two-parameter family of tranmormations depending on T, x_0. Define the set

(7.11) $$\mathcal{G} = \{(T, x_0); e^{-1-\hat{\vartheta}} \leq T \leq e^{1-\hat{\vartheta}}, |\hat{x} - x_0| \leq e^{2(1-\hat{\vartheta})/3}\}.$$

Elements of $\mathcal{G}$ will be our "guesses" for the exact time and location of the blow-up. For each fixed $(T, x_0) \in \mathcal{G}$, call $\xi(\tau)$ the point at which the rescaled variables $S(\tau, \cdot)$ attains its maximum. Assuming that $S_{\eta\eta}$ is negative at ξ, from (7.9) and the relations

(7.12) $$S_\eta(\tau, \xi(\tau)) \equiv 0, \quad S_{\eta\tau}(\tau, \xi(\tau)) + S_{\eta\eta}(\tau, \xi(\tau))\dot{\xi}(\tau) \equiv 0,$$

(7.13) $$S_{\eta\tau} + \tfrac{\eta}{2} S_{\eta\eta} + \tfrac{1}{2} S_\eta + \lambda e^{-\tau/2}(v_{\eta\eta} - w_{\eta\eta}) = e^S S_\eta,$$

one obtains

(7.14) $$\frac{d\xi}{d\tau} = -S_{\eta\tau}(\tau,\xi)/S_{\eta\eta}(\tau,\xi) = \frac{\xi(\tau)}{2} + \frac{\lambda}{S_{\eta\eta}}e^{-\tau/2}(v_{\eta\eta} - w_{\eta\eta}),$$

(7.15) $$\frac{d}{d\tau}S(\tau,\xi(\tau)) = S_\tau + S_\eta\dot\xi = e^{S(\tau,\xi)} - 1 - \lambda e^{-\tau/2}(v_\eta(\tau,\xi) - w_\eta(\tau,\xi)).$$

The initial conditions for $\xi, S(\xi)$ are

(7.16) $$\xi(\tau_0) = (\hat{x} - x_0)T^{-1/2}, \quad S(\xi(\tau_0)) = \hat\vartheta + \ln T.$$

In $\mathbf{R}^3$, define the tube with shrinking square section

(7.17) $$\mathcal{T} = \{(\tau,\xi,S); \quad |\xi|,|S| \leq e^{-\tau/3}\}.$$

As (T,x_0) range in $\mathcal{G}$, (7.14)-(7.16) determine a two-parameter family of trajectories $\tau \to (\xi(\tau), S(\tau,\xi(\tau)))$ depending continuously on T, x_0. We will show that these trajectories are well defined (in particular $S_{\eta\eta}(\tau,\xi(\tau)) < 0$ as long as they remains inside $\mathcal{T}$. Moreover, all boundary points of $\mathcal{T}$ are strict exit points. A topological argument thus implies the existence of some $T, x_0 \in \mathcal{G}$ whose corresponding solution of (7.14)-(7.16) remains forever inside $\mathcal{T}$. Such T and x_0 provide the exact timing and location of the blow-up. In the second part of the proof we refine the estimates on u,v,w and on their derivatives, and establish the limits (7.1) by means of a comparison theorem.

8 Proof of Theorem 7.2: step 1

For $(T,x_0) \in \mathcal{G}$, let u,v,w,S be the corresponding solution of (7.8)-(7.10), defined for $\tau \geq \tau_0 = -\ln T$. Introduce the scalar quantities

$$\mu(\tau) = \max\{|u_\eta(\tau,\eta)|, |v_\eta(\tau,\eta)|, |w_\eta(\tau,\eta)|; \; |\eta| \leq e^{-\tau/6}\}$$

(8.1) $$\begin{cases} m^u(\tau) = \min u_{\eta\eta}(\tau,\eta) \\ m^v(\tau) = \min v_{\eta\eta}(\tau,\eta) \\ m^w(\tau) = \min w_{\eta\eta}(\tau,\eta) \end{cases} \quad \begin{cases} M^u(\tau) = \max u_{\eta\eta}(\tau,\eta) \\ M^v(\tau) = \max v_{\eta\eta}(\tau,\eta) \\ M^w(\tau) = \max w_{\eta\eta}(\tau,\eta), \end{cases}$$

(8.2) $$D(\tau) = \max\{|u_{\eta\eta}(\tau,\eta)|, |v_{\eta\eta}(\tau,\eta)|, |w_{\eta\eta}(\tau,\eta)|\}.$$

where the min and max are all taken over the set $|\eta| \leq e^{-\tau/6}$. Defining

(8.3) $$m = m^u + m^v + m^w, \quad M = M^u + M^v + M^w,$$

we clearly have

(8.4) $$m(\tau) \leq S_{\eta\eta}(\tau,\eta) \leq M(\tau)$$

whenever $|\eta| \leq e^{-\tau/6}$. Estimates on the initial values of μ, m, M and D can be derived from the assumptions i)-iii) in Theorem 7.2. Recalling that

(8.5) $$\hat{\vartheta} + 1 \geq \tau_0 = -\ln T \geq \hat{\vartheta} - 1 > \tilde{\tau} > 3,$$

if $|\eta| \leq e^{-\tau_0/6}$, then

$$|x - x_0| = |\eta|e^{-\tau_0/2} \leq e^{-2\tau_0/3} \leq e^{2(1-\hat{\vartheta})/3},$$

$$|x - \hat{x}| \leq |x - x_0| + |x_0 - \hat{x}| \leq 2e^{2(1-\hat{\vartheta})/3},$$

The hypotheses i)-iii) therefore imply

(8.6) $$\begin{cases} \mu(\tau_0) < (\gamma/2)e^{\hat{\vartheta}-1/2} \cdot e^{-\tau_0/2} \leq \gamma/2 \\ m(\tau_0) > [-e^{\hat{\vartheta}-1} \cdot \beta/2] \cdot e^{-\tau_0} \geq -\beta/2 \\ M(\tau_0) < [-2\alpha e^{\hat{\vartheta}+1}]e^{-\tau_0} \leq -2\alpha \\ D(\tau_0) < [e^{\hat{\vartheta}-1}\beta/2]e^{-\tau_0} \leq \beta/2. \end{cases}$$

To estimate μ, m, M, D when $\tau > \tau_0$, we differentiate (7.8) and obtain

(8.7) $$\begin{cases} u_{\eta\tau} + \frac{\eta}{2}u_{\eta\eta} + \frac{u_\eta}{2} = Ae^S S_\eta \\ v_{\eta\tau} + \frac{\eta}{2}v_{\eta\eta} + \frac{v_\eta}{2} = Be^S S_\eta - \lambda e^{-\tau/2}v_{\eta\eta} \\ w_{\eta\tau} + \frac{\eta}{2}w_{\eta\eta} + \frac{w_\eta}{2} = Be^S S_\eta + \lambda e^{-\tau/2}w_{\eta\eta}, \end{cases}$$

(8.8) $$\begin{cases} u_{\eta\eta\tau} + u_{\eta\eta} + \frac{\eta}{2}u_{\eta\eta\eta} = Ae^S(S_\eta^2 + S_{\eta\eta}) \\ v_{\eta\eta\tau} + v_{\eta\eta} + \frac{\eta}{2}v_{\eta\eta\eta} = Be^S(S_\eta^2 + S_{\eta\eta}) - \lambda e^{-\tau/2}v_{\eta\eta\eta} \\ w_{\eta\eta\tau} + w_{\eta\eta} + \frac{\eta}{2}w_{\eta\eta\eta} = Be^S(S_\eta^2 + S_{\eta\eta}) + \lambda e^{-\tau/2}w_{\eta\eta\eta}. \end{cases}$$

Observe that the three families of characteristics for the system (7.8) are determined by the equations

$$\dot{\eta} = \frac{\eta}{2}, \quad \dot{\eta} = \frac{\eta}{2} + \lambda e^{-\tau/2}, \quad \dot{\eta} = \frac{\eta}{2} - \lambda e^{-\tau/2}.$$

By (8.5) and the assumption C1) on $\tilde{\tau}$, when $\tau \geq \tilde{\tau}$ all characteristics are flowing out from the domain $\{(\tau, \eta); |\eta| \leq e^{-\tau/6}\}$. Assume that $(\tau, \xi(\tau), S(\xi(\tau))) \in \mathcal{T}$ and $m(\tau) \leq M(\tau) < 0$ for τ in some initial interval $[\tau_0, \tau')$. Then the following estimates hold.

(8.9) $$S(\tau, \eta) \leq e^{-\tau/3}, \quad e^{S(\tau,\eta)} \leq 1 + 2e^{-\tau/3} \leq 2,$$

(8.10) $$|S_\eta(\tau,\eta)| \leq |\eta - \xi(\tau)||m(\tau)| \leq 2|m(\tau)|e^{-\tau/6}.$$

Using (8.9),(8.10), from (8.7),(8.8) we deduce

(8.11)
$$\dot{\mu}(\tau) \leq -\mu(\tau)/2 + \max\{|S_\eta(\tau,\eta)| \cdot e^{S(\tau,\eta)}; |\eta| \leq e^{-\tau/6}\}$$
$$\leq -\mu(\tau)/2 + 4|m(\tau)|e^{-\tau/6},$$

(8.12)
$$\begin{cases} \dot{m}^u(\tau) \geq -m^u(\tau) + Am(\tau)(1 + 2e^{-\tau/3}) \\ \dot{m}^v(\tau) \geq -m^v(\tau) + Bm(\tau)(1 + 2e^{-\tau/3}) \\ \dot{m}^w(\tau) \geq -m^w(\tau) + Bm(\tau)(1 + 2e^{-\tau/3}), \end{cases}$$

$$\dot{m}(\tau) \geq 2e^{-\tau/3}m(\tau),$$

(8.13)
$$\begin{cases} \dot{M}^u(\tau) \leq -M^u(\tau) + 2A(2|m(\tau)|e^{-\tau/6})^2 + A(1 + 2e^{-\tau/3})M(\tau) \\ \dot{M}^v(\tau) \leq -M^v(\tau) + 2B(2|m(\tau)|e^{-\tau/6})^2 + B(1 + 2e^{-\tau/3})M(\tau) \\ \dot{M}^w(\tau) \leq -M^w(\tau) + 2B(2|m(\tau)|e^{-\tau/6})^2 + B(1 + 2e^{-\tau/3})M(\tau) \end{cases}$$

$$\dot{M}(\tau) \leq 8m^2(\tau)e^{-\tau/3}.$$

The assumption C1) on $\tilde{\tau}$ now implies

(8.14) $$\mu(\tau) < \gamma, \quad m(\tau) > -\beta, \quad M(\tau) < -\alpha$$

for all $\tau \in [\tau_0, \tau')$. Since $S_{\eta\eta} < M(\tau) < 0$, (8.14) shows that solutions (ξ, S) of (7.14),(7.15) are well defined and depend continuously on the parameters T, x_0, as long as (ξ, S) remain inside the tube $\mathcal{T}$. Moreover, from (8.8) it follows

(8.15)
$$\dot{D}(\tau) \leq -D(\tau) + 2\max\{(S_\eta^2 + |S_{\eta\eta}|); |\eta| \leq e^{-\tau/6}\}$$
$$\leq -D(\tau) + 2[(2|m(\tau)|e^{-\tau/6})^2 + |m(\tau)|]$$
$$\leq -D(\tau) + 8\beta^2 e^{-\tau/3} + 2\beta,$$

hence, by condition C2),

(8.16) $$D(\tau) < 3\beta \quad \forall \tau \in [\tau_0, \tau').$$

We now define a continuous map φ from the unit square $Q = \{(y_1, y_2); |y_1|, |y_2| \leq 1\}$ to its boundary as follows. Given $(y_1, y_2) \in Q$, there exists a unique $(T, x_0) \in \mathcal{G}$ such that, setting $\tau_0 = -\ln T$, one has

(8.17)
$$(y_1, y_2) = e^{\tau_0/3}(\xi(\tau_0), S(\xi(\tau_0)))$$
$$= T^{-\frac{1}{3}}((\hat{x} - x_0)T^{-\frac{1}{2}}, \hat{\vartheta} + \ln T).$$

If the corresponding solution of (7.14)-(7.16) hits the boundary of T at the first time τ, define

(8.18) $\qquad \varphi(y_1, y_2) = e^{\tau/3}(\xi(\tau), S(\tau, \xi(\tau))) \in \partial Q.$

As customary, the continuity of φ can be proved by showing that every boundary point of T with $\tau \geq \tilde{\tau}$ is a strict exit point for solution of (7.14)-(7.16). Indeed, if $|\xi(\tau)| = e^{-\tau/3}$, using (8.16) in (7.14) one obtains

$$\frac{d}{d\tau}|\xi| \geq \frac{1}{2}e^{-\tau/3} - \frac{\lambda}{|M(\tau)|}e^{-\tau/2} \cdot 2D(\tau)$$

$$\geq \frac{1}{2}e^{-\tau/3} - \frac{6\beta\lambda}{\alpha}e^{-\tau/2} > 0$$

because of C1). On the other hand, if $|S(\tau, \xi(\tau))| = e^{-\tau/3}$, using (8.14) in (7.15) we now have

$$\frac{d}{d\tau}|S(\tau, \xi(\tau))| \geq \frac{1}{2}e^{-\tau/3} - \lambda e^{-\tau/2} \cdot 2\mu(\tau) > \frac{1}{2}e^{-\tau/3} - 2\lambda\gamma e^{-\tau/2} > 0.$$

If every choice $(T, x_0) \in \mathcal{G}$ yields a trajectory which escapes from T at some finite time τ, the map φ would then be a continuous surjection from Q onto ∂Q which leaves the points of ∂Q fixed. Since no such map exists, we have proved the existence of some $(T, x_0) \in \mathcal{G}$ whose corresponding trajectory $\tau \to (\xi(\tau), S(\tau, \xi(\tau)))$ remains inside T for all times $\tau \geq \tau_0$.

9 Proof of Theorem 7.2: step 2

From now on, everything is referred to a unique coordinate tranmormation, i.e., the one determined by the element $(T, x_0) \in \mathcal{G}$ singled out at the end of §8. Differentiating (8.8) once again one finds

(9.1) $\begin{cases} u_{\eta\eta\eta\tau} + \frac{3}{2}u_{\eta\eta\eta} + \frac{\eta}{2}u_{\eta\eta\eta\eta} = Ae^S S_{\eta\eta\eta} + Ae^S S_\eta(S_\eta^2 + 3S_{\eta\eta}) \\ v_{\eta\eta\eta\tau} + \frac{3}{2}v_{\eta\eta\eta} + (\frac{\eta}{2} + \lambda e^{-\tau/2})v_{\eta\eta\eta\eta} = Be^S S_{\eta\eta\eta} + Be^S S_\eta(S_\eta^2 + 3S_{\eta\eta}) \\ w_{\eta\eta\eta\tau} + \frac{3}{2}w_{\eta\eta\eta} + (\frac{\eta}{2} - \lambda e^{-\tau/2})w_{\eta\eta\eta\eta} = Be^S S_{\eta\eta\eta} + Be^S S_\eta(S_\eta^2 + 3S_{\eta\eta}). \end{cases}$

Call $E^u(\tau)$ the maximum of $|u_{\eta\eta\eta}(\tau, \cdot)|$ over the set $|\eta| \leq e^{-\tau/6}$, and define $E^v(\tau)$, $E^w(\tau)$ similarly. Set $E = E^u + E^v + E^w$. Since $|S_{\eta\eta\eta}(\tau, \eta)| \leq E(\tau)$, by (9.1) E satisfies the differential inequality

(9.2) $\qquad \dot{E}(\tau) \leq (-\frac{3}{2} + \max_\eta e^{S(\tau,\eta)})E(\tau) + \max_\eta |S_\eta| \cdot |e^S S_\eta^2 + 3e^S S_{\eta\eta}|.$

From (8.10) and (8.14) it follows

(9.3) $\qquad \lim_{\tau \to \infty} \max\{|S_\eta(\tau, \eta)|;\ |\eta| \leq e^{-\tau/6}\} = 0.$

Using (9.3) and the uniform bounds on $S, |S_{\eta\eta}|$ in (9.2) we obtain
$$\lim_{\tau \to \infty} E(\tau) = 0,$$
therefore recalling the definitions (8.3),

(9.4) $$\lim_{\tau \to \infty} |M(\tau) - m(\tau)| = 0.$$

By (8.4), (8.14), from (9.4) it follows that

(9.5) $$\Omega = \lim_{\substack{\tau \to \infty \\ |\eta| \le e^{-\tau/6}}} S_{\eta\eta}(\tau, \eta)$$

exists and satisfies

(9.6) $$-\beta \le \Omega \le -\alpha.$$

Using (9.5) in (8.8) we now obtain

(9.7) $$\begin{cases} \lim_{\tau \to \infty} u_{\eta\eta} = A\Omega, \\ \lim_{\tau \to \infty} v_{\eta\eta} = \lim_{\tau \to \infty} w_{\eta\eta} = B\Omega, \end{cases}$$

all limits being taken inside the region $|\eta| \le e^{-\tau/6}$. The equations (7.14),(7.15) in view of (9.5),(9.7) yield

(9.8) $$|\xi(\tau)| \le C_1 e^{-\tau/2}, \quad |S(\tau, \xi(\tau))| \le C_1 e^{-\tau/2},$$

for some constant C_1 and all τ large enough. Since $S_\eta(\tau, \xi(\tau)) \equiv 0$ and $|S_{\eta\eta}| < \beta$, (9.8) implies an estimate of the form

(9.9) $$|S_\eta(\tau, \xi)| \le C_2 e^{-\tau/2}$$

whenever $|\eta| \le (C_1 + 2\lambda) e^{-\tau/2}$. In particular, (9.8) and (9.9) yield

(9.10) $$|S(\tau, 0)| \le C_3 e^{-\tau/2},$$

valid for some constant C_3 and all τ large enough. Define

$$\mu'(\tau) = \max\{|u_\eta(\tau, \eta)|, |v_\eta(\tau, \eta)|, |w_\eta(\tau, \eta)|;\ |\eta| \le 2\lambda e^{-\tau/2}\}.$$

An estimate entirely similar to (8.11) now yields

$$\begin{aligned} \mu'(\tau) &\le -\mu'(\tau)/2 + \max\{|S_\eta(\tau, \eta)| e^{S(\tau, \eta)};\ |\eta| \le 2\lambda e^{-\tau/2}\} \\ &\le -\mu'(\tau)/2 + 2C_2 e^{-\tau/2}, \end{aligned}$$

which implies, in particular,

(9.11) $$|u_\eta(\tau, 0)|, |v_\eta(\tau, 0)|, |w_\eta(\tau, 0)| \le C_4 \tau e^{-\tau/2}.$$

To extend our estimates beyond the narrow strip $|\eta| \le 2\lambda e^{-\tau/2}$ we rely on a comparison technique.

Lemma 9.1 Let (u, v, w) be a solution of (7.8) and let $Z = Z(\tau, \eta)$ be a scalar function such that
(9.12) $$Z_\tau + \tfrac{\eta}{2} Z_\eta = e^Z - 1.$$
Call
$$u^-(\tau, \eta) = \min\{u(\tau, \eta'); |\eta' - \eta| \leq 2\lambda e^{-\tau/2}\},$$
$$u^+(\tau, \eta) = \max\{u(\tau, \eta'); |\eta' - \eta| \leq 2\lambda e^{-\tau/2}\},$$
and similarly for v^-, v^+, w^-, w^+. If at some point (τ, η) we have $(u^- + v^- + w^-)(\tau, \eta) \geq Z(\tau, \eta)$, then
(9.13) $$(u^- + v^- + w^-)(\tau + t, \eta e^{t/2}) \geq Z(\tau + t, \eta e^{t/2}) \quad \forall t > 0.$$
On the other hand, if $(u^+ + v^+ + w^+)(\tau, \eta) \leq Z(\tau, \eta)$, then
(9.14) $$(u^+ + v^+ + w^+)(\tau + t, \eta e^{t/2}) \leq Z(\tau + t, \eta e^{t/2}) \quad \forall t > 0.$$

Indeed, (9.13) and (9.14) can be proved by defining
$$Y^\pm(t) = (u^\pm + v^\pm + w^\pm)(\tau + t, \eta e^{t/2})$$
and checking that
$$\frac{dY^-(t)}{dt} \geq \frac{d}{dt} Z(\tau + t, \eta e^{t/2}),$$

$$\frac{dY^+(t)}{dt} \leq \frac{d}{dt} Z(\tau + t, \eta e^{t/2})$$
whenever $Y^- \geq Z$ or $Y^+ \leq Z$ respectively.

Using Lemma 9.1, we now prove that as $\tau \to \infty$ the functions u, v, w, S converge to some limit, uniformly on bounded sets. Fix $\varepsilon > 0$ and define
$$Z^+(\tau, \eta) = -\ln(1 - \frac{\Omega + \varepsilon}{2} \eta^2),$$
$$Z^-(\tau, \eta) = -\ln(1 - \frac{\Omega - \varepsilon}{2} \eta^2).$$
Observe that Z^+, Z^- are time-invariant solutions of (9.12), with $Z^+_{\eta\eta}(0) = \Omega + \varepsilon$, $Z^-_{\eta\eta}(0) = \Omega - \varepsilon$. For any $\varepsilon' > 0$, by (9.7) there exists τ_1 so large that

(9.15) $$\begin{cases} u(\tau, \eta) \leq u(\tau, 0) + u_\eta(\tau, 0) \cdot \eta + (A\Omega + \varepsilon')\eta^2/2 \\ v(\tau, \eta) \leq v(\tau, 0) + v_\eta(\tau, 0) \cdot \eta + (B\Omega + \varepsilon')\eta^2/2 \\ w(\tau, \eta) \leq w(\tau, 0) + w_\eta(\tau, 0) \cdot \eta + (B\Omega + \varepsilon')\eta^2/2 \end{cases}$$

whenever $|\eta| \leq e^{-\tau/6}$, $\tau \geq \tau_1$.

Call $u_\infty, v_\infty, w_\infty$ respectively the limits of $u(\tau,0), v(\tau,0), w(\tau,0)$ as $\tau \to \infty$. Using (9.10), (9.11) in (7.8), it is clear that these limits exist. In fact, we have the estimate

(9.16) $\quad |u(\tau,0) - u_\infty|, \ |v(\tau,0) - v_\infty|, \ |w(\tau,0) - w_\infty| \leq C_5 \tau e^{-\tau/2}$

for some constant C_5 and all τ large enough. Moreover,

$$u_\infty + v_\infty + w_\infty = \lim_{\tau \to \infty} S(\tau,0) = 0.$$

By (9.15),(9.16), there exists $\bar{\tau}$ so large that

(9.17) $\quad u^+ - u_\infty \leq AZ^+, \quad v^+ - v_\infty \leq BZ^+, \quad w^+ - w_\infty \leq BZ^+$

at every point of the set $\{(\tau,\eta); \tau \geq \bar{\tau}, |\eta| = \frac{1}{2}e^{-\tau/6}\}$. In particular,

(9.18) $\quad (u^+ + v^+ + w^+)(\tau, \pm e^{-\tau/6}) \leq Z^+(\tau, \pm\frac{1}{2}e^{-\tau/6})$

for all $\tau \geq \bar{\tau}$. The previous Lemma now yields

(9.19) $\quad S(\tau,\eta) \leq (u^+ + v^+ + w^+)(\tau,\eta) \leq Z^+(\tau,\eta)$

on the region

$$\Sigma_{\bar{\tau}} = \{(\tau,\eta); \ \tau \geq \bar{\tau}, \ \tfrac{1}{2}e^{-\tau/6} \leq |\eta| \leq \tfrac{1}{2}e^{(\tau-\bar{\tau})/2} \cdot e^{-\tau/6}\}.$$

For $\bar{\tau}$ suitably large, an entirely similar argument yields

(9.20) $\quad S(\tau,\eta) \geq (u^- + v^- + w^-)(\tau,\eta) \geq Z^-(\tau,\eta)$

for all $(\tau,\eta) \in \Sigma_{\bar{\tau}}$. Since $S(\tau,\eta) \to 0$ as $\tau \to \infty$ on the strip $|\eta| \leq \frac{1}{2}e^{-\tau/6}$, (9.19) and (9.20) prove the following. For every $\varepsilon > 0$ there exists $\bar{\tau}$ large enough so that

(9.21) $\quad |S(\tau,\eta) + \ln(1 - \tfrac{\Omega}{2}\eta^2)| < \varepsilon$

for every (τ,η) in the region

$$R_{\bar{\tau}} = \{(\tau,\eta); \ \tau \geq \bar{\tau}, \ |\eta| \leq \frac{1}{2}e^{\tau/2}e^{-2\bar{\tau}/3}\}.$$

Using (9.19) and (9.20) in (7.8), another comparison argument shows that on $\Sigma_{\bar{\tau}}$ the functions $u^\pm$ satisfy

$$AZ^-(\eta) \leq u^-(\tau,\eta) - u_\infty \leq u^+(\tau,\eta) - u_\infty \leq AZ^+(\eta),$$

while the functions $v^\pm - v_\infty$ and $w^\pm - w_\infty$ are bounded by $BZ^\pm$. On the strip $|\eta| \leq \frac{1}{2}e^{-\tau/6}$, estimates on u,v,w are already known. Therefore we conclude that for every $\varepsilon > 0$ there exists $\bar\tau$ such that

(9.22)
$$|u(\tau,\eta) - u_\infty + A\ln(1 - \tfrac{\Omega}{2}\eta^2)| < \varepsilon,$$
$$|v(\tau,\eta) - v_\infty + B\ln(1 - \tfrac{\Omega}{2}\eta^2)| < \varepsilon,$$
$$|w(\tau,\eta) - w_\infty + B\ln(1 - \tfrac{\Omega}{2}\eta^2)| < \varepsilon,$$

for every $(\tau,\eta) \in R_{\bar\tau}$. Setting $z_1^\infty = u_\infty, z_2^\infty = v_\infty, z_3^\infty = w_\infty$, the reinterpretation of (9.22) in the original variables t,x,z is the following. For every $\varepsilon > 0$ there exists $\bar t < T$ such that the estimates (7.1) hold whenever $\bar t \leq t < T$, $|x - x_0| \leq \frac{1}{2}(T-\bar t)^{2/3}$. This proves theorem 7.2.

The statements in Theorem 7.1 now follow as corollaries. Indeed, for any given $\alpha, \beta, \gamma, \bar\tau$, the hypotheses in Theorem 7.2 are satisfied by an open set of functions. The previous proof also indicates that the parameters T, x_0, z_i^∞ and Ω, which characterize the self-similar blow-up, depend continuously on the initial data $\bar z$ in the C^2 topology.

References

[1] Bebernes, J. and Bressan, A. (1982). Thermal behavior for a confined reactive gas, *Journal of Differential Equations* **44**, 118-33.

[2] Bebernes, J., Bressan, A. and Eberly, D. (1987). A description of blowup for the solid fuel ignition model, *Indiana University Mathematics Journal*, **36**, 295-305.

[3] Bebernes, J. and Kassoy, D. (1989). Reactive-Euler induction models, *Proceedings of the Sixth Army Conference on Applied Mathematics & Computing*, ARO Report 89-1, 473-82.

[4] Bebernes, J. and Kassoy, D. (1988). Characterizing self-similar blowup, *Mathematical Modeling in Combustion and Related Topics*, **NATO ASI** Series, 383-92.

[5] Bressan, A. Blowup asymptotics for the reactive-Euler model, *SIAM Journal of Mathematical Analysis*, in press.

[6] Clarke, J. F. and Cant, R. S. (1985). Nonsteady gasdynamic effects in the induction domain behind a strong shock wave, *Dynamics of Flames and Reactive Systems, Progress in Aeronautics and Astronautics*, **95**, 142-63.

[7] Courant, R. and Hilbert, D. (1962). *Methods of mathematical physics*, Vol. **II**, Wiley Interscience Pub., New York.

[8] Dold, J. W. (1988). Dynamics transition of a self-igniting region, *Mathematical Modeling in Combustion and Related Topics*. C.-M. Brauner and C. Schmidt-Lainé, Eds., **NATO ASI** series, M. Nijhoff, Pub. 461-70.

[9] Giga, Y. and Kohn, R. (1985). Asymptotically self-similar blow up of semilinear heat equations, *Communications on Pure and Applied Mathematics*, **38**, 297-319.

[10] Hartman, P. (1982). *Ordering Differential Equations*, Birkhäuser, Boston.

[11] Jackson, T. L. and Kapila, A. K. (1985). Shock-induced thermal runaway, *SIAM Journal of Applied Mathematics*, **45**, 130-37.

[12] Jackson, T. L., Kapila, A. K. and Stewart, D. S. (1989). Evolution of a reaction center in an explosive material, *SIAM Journal of Applied Mathematics*, **49**, 432-58.

[13] Kassoy, D. and Poland, J. (1983). The induction period of a thermal explosion in a gas between infinite parallel plates, *Combustion and Flame*, **50**, 259-74.

[14] Kassoy, D. R., Kapila, A. K. and Stewart, D. S. (1989). A unified formulation for diffusive and nondiffusive thermal explosion theory, *Combustion Science and Technology*, **63**, 33-43.

[15] Kassoy, D., Riley, N., Bebernes, J. and Bressan, A. (1989). The confined nondiffusive thermal explosion with spatially homogeneous pressure variation, *Combustion Science and Technology*, **63**, 45-62.

[16] Martin, R. H. Jr. (1976). *Nonlinear operators and differential equations in Banach spaces*, Wiley, New York.

A new scaling of a problem in combustion theory

G. C. Wake
Massey University

J. B. Burnell
DSIR, Wellington

J. G. Graham-Eagle
University of Auckland

B. F. Gray
Macquarie University

1 Introduction

The well-established approach to the theory of thermal ignition expresses the heat balance equation in terms of the dimensionless variables introduced by Frank-Kamenetskii(1955). Unfortunately, this choice of scaling confuses the role of ambient temperature to such an extent that it is almost impossible to consider the effect of varying the ambient temperature alone. The purpose of this paper is to use different dimensionless variables which allow the ambient temperature to be varied independently of the remaining system parameters. The problem obtained from this approach is nonstandard in as much as the bifurcation parameter occurs in the boundary condition, but it is nevertheless amenable to treatment by the usual methods of functional analysis; see Burnell et al. (1989) for a summary of results.

The equation for the heat balance in the original variables is, neglecting reactant consumption,

$$\nabla \cdot (k \nabla T) + \sigma Q A \exp(-E/RT) = C \frac{\partial T}{\partial t} \text{ in } \Omega.$$

Here Ω is the bounded domain in $\mathbf{R}^n$ with smooth boundary $\partial \Omega$ in which the reaction is taking place, Q is the exothermicity, σ the density, $A \exp(-E/RT)$ the dependence of the reaction rate on the temperature T (measured in $^o K$) derived from the Arrhenius law, k the thermal conductivity, E the activation energy, R the gas constant, and C the thermal capacity. For simplicity all these (except T) are presumed constant.

In this discussion the temperature on the boundary is held at the fixed value T_a, although the theory can be extended, with few significant changes, to the case

$$k \frac{\partial T}{\partial n} + h(T - T_a) = 0 \text{ on } \partial \Omega,$$

in which h is the heat transfer coefficient and $\partial/\partial n$ is the outward normal derivative on $\partial\Omega$.

The steady state formulation of this problem is due essentially to Frank-Kamenetskii(1955) and has been developed by many authors; see for example the reviews by Gray and Lee(1967), Gray and Sherrington(1977), and Boddington, Gray and Wake(1977). These analyses traditionally use a dimensionless temperature rise over the ambient given by

$$\theta = \frac{E}{RT_a^2}(T - T_a)$$

and so the equation become

$$\nabla^2\theta + \delta\exp[\theta/(1+\varepsilon\theta)] = 0 \text{ in } \Omega, \theta = 0 \text{ on } \partial\Omega .$$

Here $\varepsilon = RT_a/E$ is considered a constant and the Frank-Kamenetskii parameter

$$\delta = \frac{\sigma Q a_0^2 E A \exp(-E/RT_a)}{kRT_a^2}$$

has the role of dimensionless eigenparameter, where $2a_0$ is some characteristic length scale for the problem. Typically $E/RT_a = \varepsilon^{-1}$ is in the range 10-100, but smaller values are possible. This means that the nonlinearity $\exp[\theta/(1+\varepsilon\theta)]$ is convex for small positive θ and concave for large θ.

This and related problems have been extensively studied, both mathematically and computationly. Such studies generally consider the behaviour of the solutions θ as the eigenvalue δ changes for fixed ε, and often focus attention on the critical value of δ at which the minimal branch of solutions is discontinuous.

From a practical viewpoint the most appropriate eigenparameter is the ambient temperature T_a and the corresponding response function is the reactant temperature T. In this case the bifurcation point in the minimal branch occurs at a critical value of T_a. However there is no simple way to obtain this critical value from a study of the problem in the Frank-Kamenetskii variables–clearly a classical (δ,θ) plot for fixed ε cannot be used to define a critical ambient temperature since the choice of ε fixes T_a for a given chemistry! The purpose of this discussion is to introduce more appropriate dimensionless variables which allow the direct use of T_a as the bifurcation parameter. Although this new formulation is in many ways equivalent to the old, there are some important distinctions.

In terms of
$$u = \frac{RT}{E} \text{ and } U = \frac{RT_a}{E}$$

the heat balance equation becomes

(1.1)
$$\nabla^2 u + \lambda\exp(-1/u) = 0 \quad \text{in } \Omega ,$$

$$u = U \quad \text{on } \partial\Omega ,$$

where
$$\lambda = \frac{\sigma Q A R a_0^2}{kE}.$$
Notice that U is the only variable or parameter that depends on T_a. Of interest in this formulation are the values of U and λ at which the solution u of (1.1) ceases to be determined as a smooth function of U and λ. Such points are often associated with the onset of thermal runaway. In this study we develop results that show the structure of the bifurcation diagram, thus providing a graphical representation of the solution set of (1.1). As already indicated, many of these results are analogous to many developed for the classical problem. However there are some marked differences. In particular (1.1) has nontrivial solutions when $U = 0$ (corresponding to absolute zero) for large λ, a phenomenon completely absent in previous studies of the problem because of the Frank-Kamenetskii scaling by ambient temperature. It should be noted that this phenomenon is independent of the heat-generation function - there is nothing special about Arrhenius here.

If the reaction is uniform(well stirred), the problem reduces to the transcendental equation
$$(u - U) = \lambda \exp(-1/u).$$
This simplified form of the problem was studied by Gray and Wake(1988) and the results are summarised in Figure 1 .

Of particular interest are the points $\lambda_{tr} = e^2/4$ at which multiplicity first occurs, and $\lambda' = e$ at which nontrivial solutions first exist when $U = 0$. These values are investigated for the general case in §2 and §3. From a practical viewpoint the value of U_{cr} is also important and is also studied in these sections. §4 is devoted to explicit calculation of the bifurcation diagram when Ω is a ball.

We end this section with some simple observations. The existence of solutions to (1.1) for all positive values of the parameters λ and U follows from a straightforward application of upper and lower solution methods. Indeed , writing w for the unique solution of the problem

(1.2) $\qquad \nabla^2 w + 1 = 0 \quad \text{in } \Omega, \quad w = 0 \quad \text{on } \partial\Omega ,$

it is simple to verify that U is a lower solution for (1.1), $U + \lambda w$ an upper solution, and (1.1) has a minimal and a maximal solution between U and $U + \lambda w$. Moreover, by the maximum principle, $U \leq u \leq U + \lambda w$ for any solution of (1.1). A more precise description of the solution for large U can be obtained from

Proposition 1 *If $\lambda > 0$, $U > 0$ and u is any solution of (1.1), then*
$$u > U + \lambda \exp(-1/U) w \text{ in } \Omega .$$

Proof. Since $u < U$ in Ω,
$$\nabla^2 (u - [U + \lambda \exp(-1/U) w]) = -\lambda [\exp(-1/u) - \exp(-1/U)] < 0 \text{ in } \Omega ,$$
$$u - [U - \lambda \exp(-1/U) w] = 0 \text{ on } \partial\Omega .$$
The result follows from the maximum principle. $\hfill$ QED

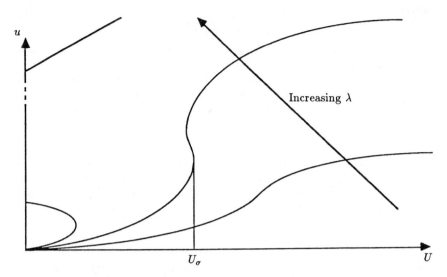

Figure 1. Bifurcation diagram of u vs U for various λ in the uniform case.

2 Multiplicity of solutions

In this section standard methods are employed for showing the uniqueness and multiplicity of solutions. In particular they rely on the maximum principle and results concerning upper and lower solutions.

We begin by showing that equation (1.1) has a unique solution if λ is sufficiently small or U is sufficiently large. The proofs given here can be simplified, but the approach taken provides estimates on the range of the parameters λ and U for which uniqueness holds.

Proposition 2 *If λ is sufficiently small then (1.1) has a unique solution for all $U \geq 0$.*

Proof. Let μ_1 denote the principal eigenvalue of the problem

(2.1) $$\nabla^2 \varphi + \mu \varphi = 0 \text{ in } \Omega, \varphi = 0 \text{ on } \partial\Omega.$$

It is well known that $\mu_1 > 0$ and that the corresponding eigenfunction φ_1 is positive in Ω. Now let u_1 and u_2 denote the minimal and maximal solutions of (1.1) respectively. Then the function $\psi = u_2 - u_1$ is nonnegative on Ω and satisfies

(2.2) $$\nabla^2 \psi + \lambda[\exp(-1/\zeta)/\zeta^2]\psi = 0 \text{ in } \Omega,$$
$$\psi = 0 \text{ on } \partial\Omega,$$

where ζ is a function with $u_1 \leq \zeta \leq u_2$, Since $u_1 \geq 0$ and $\exp(-1/\zeta)/\zeta^2 \leq 4/e^2$ if $\zeta \geq 0$, it follows that
$$\nabla^2 \psi + \lambda \left(4/e^2\right) \psi \geq 0 \text{ in } \Omega.$$

Multiplying this inequality by φ_1, integrating over Ω and applying Green's theorem yields

$$\begin{aligned} 0 &\leq \int_\Omega \varphi_1 \nabla^2 \psi dx + 4e^{-2}\lambda \int_\Omega \varphi_1 \psi dx \\ &= [4e^{-2}\lambda - \mu_1] \int_\Omega \varphi_1 \psi dx, \end{aligned}$$

from which it is easy to see that if

(2.3) $$\lambda < e^2 \mu_1 / 4$$

then $\psi = 0$ and so (1.1) has a unique solution as required. QED

Proposition 3 *For every $U_0 > 0$ there exists λ_0 so that (1.1) has a unique solution whenever $U > U_0$ and $\lambda > \lambda_0$.*

Proof. Consider the Banach space C_w defined by

$$C_w = \{u \in C(\bar{\Omega}) : \|u\|_w = \sup_{x \in \Omega} \frac{|u(x)|}{|w(x)|} \text{ is finite}\}$$

where w is defined in (1.2). Note that if $u \in C_w$ then $u = 0$ on $\partial \Omega$. Since the inverse of $-\nabla^2$ with the boundary condition $u = 0$ is a compact map $G : L^2(\Omega) \to W_0^{2,q}(\Omega)$, which is continuously embedded in $C_0^1(\Omega)$ for $q > n$ and so in C_w, it follows that G is a compact map from $L^q(\Omega)$ into C_w if $q > n$.

Now, in the notation of Proposition 1, $\psi \in C_w$ and

$$\psi = G\left(\lambda[\exp\left(-1/\zeta\right)/\zeta^2]\psi\right)$$

so if $q > n$,

(2.4) $$\begin{aligned} \|\psi\|_w &\leq \|G\|_{w,q} \|\lambda[\exp\left(-1/\zeta\right)/\zeta^2]\psi\|_{L^q} \\ &= \|G\|_{w,q} \|\lambda[\exp\left(-1/\zeta\right)/\zeta^2]w\|_{L^q} \|\psi\|_w. \end{aligned}$$

Here $\|G\|_{w,q}$ is the operator norm of G.

We now show that

$$I = \|\lambda[\exp\left(-1/\zeta\right)/\zeta^2]w\|_{L^q}^q = \int_\Omega (\lambda[\exp\left(-1/\zeta\right)/\zeta^2]w)^q dx$$

tends to 0 as $\lambda \to \infty$. Since $\lambda w \leq e^{1/U}(\zeta - U)$ it follows that, for $U >$ fixed $U_0 > 0$,

$$I \leq \int_\Omega ([\exp(-1/\zeta)/\zeta^2]\exp(1/U)(\zeta - U))^q dx$$

$$\leq \int_\Omega ([\exp(-1/\zeta)/\zeta^2]\exp(1/U_0)(\zeta - U_0))^q dx$$

$$\leq \int_\Omega [\exp(1/U_0)(\zeta - U_0)/\zeta^2]^q dx.$$

Moreover, from Proposition 1,

$$\zeta \geq U + \lambda \exp(-1/U)w \geq U_0 + \lambda \exp(-1/U_0)w$$

so for each $x \in \Omega$, $\zeta \to \infty$ and $\exp(1/U_0)(z - U_0)/\zeta^2 \to 0$ as $\lambda \to \infty$ uniformly in U for $U > U_0$. Also, since $\zeta \geq U_0$, $[\exp(1/U_0)(\zeta - U_0)/\zeta^2]^q \leq [\exp(1/U_0)/4U_0]^q$ in Ω, and the dominated convergence theorem implies

$$\int_\Omega [\exp(1/U_0)(\zeta - U_0)/\zeta^2]^q dx \to 0$$

uniformly as $\lambda \to \infty$ for $U > U_0$. In particular it follows that there exists $\lambda_0 > 0$ so that, whenever $\lambda > \lambda_0$ and $U > U_0$,

$$I \leq \int_\Omega [\exp(1/U_0)(\zeta - U_0)/\zeta^2]^q dx < \frac{1}{\|G\|_{w,q}^q}.$$

Applying this inequality to (2.4) shows that, for $\lambda \geq \lambda_0$ and $U \geq U_0$,

$$\|\psi\|_w \leq k\|\psi\|_w$$

where k is a constant less than unity. Clearly this is satisfied only by $\psi = 0$ and the theorem is proved. QED

The proof given here follows the approach of Dancer(1986), modified to show that λ_0 can be chosen independently of U for $U \geq U_0$. The obvious consequence of this result is that as $\lambda \to \infty$, the region of values of U for which (1.1) has a unique solution increases to $(0, \infty)$.

Existence was deduced from the upper/lower solution pair $U < U + \lambda w$. To prove there is more than one solution, we construct intermediate upper and lower solutions

$$U < \text{Upper Solution} < \text{Lower Solution} < U + \lambda w.$$

Proposition 4 *If $U \exp(1/2U) \geq \lambda\|w\|_0$, then $U + \lambda \exp(-1/2U)w$ is an upper solution of (1.1).*

Proof. The function $\varphi = U + \lambda\exp(-1/2U)w$ satisfies
$$\nabla^2\varphi + \lambda\exp(-1/\varphi) = \lambda[-exp(-1/2U) + \exp(-1/\varphi)].$$
If U is small enough that
$$\lambda\|w\|_0 \leq \exp(1/2U)U,$$
then
$$\varphi = U + \lambda\exp(-1/2U)w \leq U + U = 2U.$$
Hence, for such values of U,
$$\nabla^2\varphi + \lambda\exp(-1/\varphi) \leq \lambda[-\exp(-1/2U) + \exp(-1/2U)] = 0 \quad \text{in } \Omega,$$
$$\varphi = U \quad \text{on } \partial\Omega.$$
So φ is an upper solution of (1.1). QED

Proposition 5 *Let $\alpha = (2\|w\|_0)^{-1}$; then if λ is sufficiently large, $U + \exp(\alpha w \ln \lambda) - 1$ is a lower solution of (1.1) for any $U \geq 0$.*

Proof. Write $\beta := \alpha \ln \lambda$ and put $\varphi = U + \exp(\beta w) - 1$. Then, since $\varphi \geq \exp(\beta w) - 1$,
$$\nabla^2\varphi + \lambda\exp(-1/\varphi) = [\beta^2|\nabla w|^2 + \beta\nabla^2 w]\exp(\beta w) + \lambda\exp(-1/\varphi)$$

$$\geq [\beta^2|\nabla w|^2 - \beta]\exp(\beta w) + \lambda\exp[-1/(\exp(\beta w) - 1)].$$

Now decompose Ω as the disjoint union
$$\Omega = \Omega_1 \cup \bar\Omega_2,$$
where Ω_1 and Ω_2 are open sets with the properties that
$$|\nabla w|^2 \geq k \geq 0 \quad \text{in } \Omega_1$$
and
$$w \geq l > 0 \quad \text{in } \Omega_2$$
for some k and l. Since $\alpha w \leq 1/2$ it follows that
$$\nabla^2\varphi + \lambda\exp(-1/\varphi) \geq \beta[\beta k - 1] \quad \text{in } \Omega_1$$
$$\geq -\beta\lambda^{1/2} + \lambda\exp[-1/(\exp(\beta w) - 1)] \quad \text{in } \bar\Omega_2.$$
If now λ is chosen so that $\beta k > 1$ and
$$\lambda\exp[-1/(\exp(\beta w) - 1)] > \beta\lambda^{1/2}$$
then φ is clearly a lower solution of (1.1) as required. QED

The existence of at least two solutions of (1.1) for large λ and small U now follows immediately from the inequality
$$U + \lambda\exp(-1/2U)w \leq U + \exp(\alpha w \ln \lambda) - 1 \quad \text{in } \Omega.$$

Notes:

1. Theorem E of Amann (1972) actually establishes the existence of at least three solutions of (1.1) lying between U and $U + \lambda w$ for large λ and small U from the above results.

2. The qualitative behaviour of the solution set (u, U, λ) which described above is the best that can be obtained. Indeed Proposition 3 implies that the region in which multiplicity may occur tends to $\{0\}$ as $\lambda \to \infty$.

3. The arguments given here hold also when $U = 0$. In this case the lower solution of Proposition 4 becomes unbounded as $\lambda \to \infty$, so therefore does the maximal solution. The minimal solution is $u = 0$. We believe that as $\lambda \to \infty$ all other solutions tend to 0. Certainly all our computations support this.

3 The bifurcation diagram

In this section, the above results are combined to obtain the qualitative behaviour of the solution set as U and λ are varied. Values of the parameters where marked changes in behaviour occur are estimated. We begin with a summary of the results:

1. For small λ, there is always a unique solution;

2. For any λ, the solution is unique for large values of U;

3. If λ is sufficiently large then multiple solutions exist for sufficiently small values of U. Figures 2 and 3 present this information graphically.

Three features of the solution set are worth further investigation:

(a) The value λ_{tr} of λ above which multiplicity first occurs–note that Proposition 2 implies $\lambda_{tr} \geq e^2 \mu_1/4$;

(b) The value U_{cr} of U at which the minimal branch is discontinuous ($\lambda > \lambda_{tr}$).

These two are related to the onset of the thermal ignition. Techniques employed in the Frank-Kamenetskii variables to compute these points for specific cases apply equally well here and so will not be discussed further. Some calculations of U_{cr} for the sphere are presented in Burnell et al. (1989).

(c) The value λ' of λ at which multiplicity first occurs for $U = 0$.

This feature of multiple solutions for $U = 0$ is completely lost in the traditional Frank-Kamenetskii scaling since nontrivial solutions u are transformed to $\theta = \infty$ when U, and so also T_a, is zero. Consequently its presence is unsuspected in the literature. Note however that it corresponds to surface temperatures of absolute zero, so is physically unrealistic. It is significant that for practical problems $\lambda > \lambda'$ so that extinction is not revealed by this phenomenon.

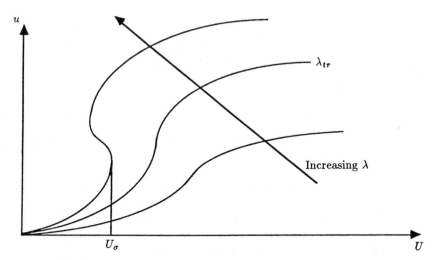

Figure 2. Bifurcation diagram for (1.1) showing λ_{tr}.

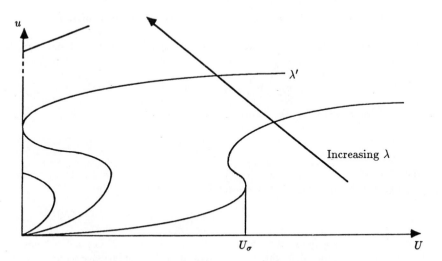

Figure 3. Bifurcation diagram for (1.1) showing λ'.

Proposition 6 *There exists λ' such that, when $U = 0$, (1.1) has a nontrivial solution if $\lambda > \lambda'$, but not if $\lambda < \lambda'$.*

Proof. It suffices to show that if (1.1) has a nontrivial solution when $U = 0$ and $\lambda = \lambda_1$, then it does so for every $\lambda > \lambda_1$. Suppose that u_1 is such a solution. Then for $\lambda > \lambda_1$ a straightforward computation shows that u_1 is a lower solution of (1.1) when $U = 0$. Since λw is an upper solution and $u_1 < \lambda_1 w < \lambda w$, it follows that (1.1) has a solution u satisfying $u_1 < u$ as required . QED

Proposition 7 $\lambda' \geq \mu_1 e$ \hspace{2em} (*cf. Proposition 2 for the definition of μ_1*)

Proof. Suppose u is a nontrivial solution of (1.1) with $U = 0$. Applying Green's theorem to the φ_1 of Proposition 2 gives

$$\begin{aligned} 0 &= \int_{\partial\Omega} \{u\frac{\partial \varphi_1}{\partial n} - \varphi_1 \frac{\partial u}{\partial n}\} dS \\ &= \int_{\Omega} [\lambda \exp(-1/u)/u - \mu_1] u \varphi_1 dx \\ &\leq \int_{\Omega} [\lambda/e - \mu_1] u \varphi_1 dx. \end{aligned}$$

Therefore $\lambda/e - \mu_1 \geq 0$ as required.

For the limiting case $\lambda = \lambda'$, it is possible to show that (1.1) has a nontrivial solution. QED

Proposition 8 *If $U = 0$ and $\lambda = \lambda'$ then (1.1) has a nontrivial solution.*

Proof. Write u_λ for the maximal solution corresponding to $U = 0$ and $\lambda = \lambda'$. Then, using the Green's function operator G of Proposition 3 we see that

$$u_\lambda - \lambda G[\exp(-1/u_\lambda)] = 0.$$

Standard arguments using the compactness of G show that

$$u' = \lim_{\lambda \to \lambda'} u_\lambda$$

exists, and u' satisfies (1.1) when $\lambda = \lambda'$. It remains to show that u' is nontrivial.

Applying the implicit function theorem to the equation

$$u - \lambda G[\exp(-1/u)] = 0$$

at the point $(u, \lambda) = (0, \lambda')$ it follows, since μ_1 (Proposition 2) is positive, that for λ in a neighbourhood of λ' there is a unique branch of solution emanating from $(0, \lambda')$. Clearly this is the trivial branch. Consequently the solutions (u_λ, λ) cannot tend to $(0, \lambda')$ as $\lambda \to \lambda'$. QED

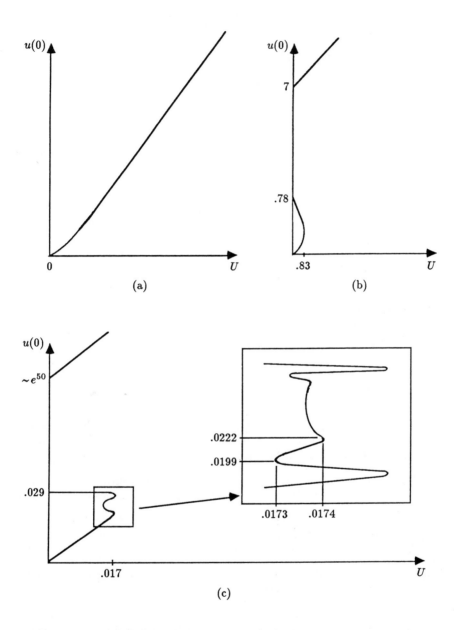

Figure 4. a, b, c. Bifurcation diagrams for (4.1) when $n = 3$ for various λ.

4   The special case of the unit ball

In the special case that Ω is the unit ball an appeal to symmetry reduces (1.1) to the ordinary differential equation

(4.1)
$$\frac{d^2u}{d^2r} + \frac{n-1}{r}\frac{du}{dr} + \lambda\exp(-1/u) = 0 \quad \text{for } 0 < r < 1,$$
$$\frac{du}{dr}(0) = u(1) = 0,$$

where r measures distance from the centre of the ball. The equivalent equation in the Frank-Kamenetskii variables $(\delta, \varepsilon, \theta)$ has been extensively studied and from this information some of the behaviour of the solution set in the variables (U, λ, u) can be obtained, though the transformation process is convoluted. Certainly, most of the features of the $(\delta, \varepsilon, \theta)$ diagrams also hold in the (U, λ, u) variables. For example, there are an arbitrarily large number of solutions for suitably chosen values of δ and ε when $n > 2$. Therefore the same will also be true in the (U, λ, u) variables. From such considerations, together with numerical computations, the following bifurcation diagrams can be found.

References

[1] Amann, H. (1972). On the number of solutions of nonlinear equations in ordered Banach Spaces. *Journal of Functional Analysis*, 11, 346-83.

[2] Boddington, T., Gray, P., and Wake, G. C. (1977). Criteria for thermal explosions with and without reactant consumption. *Proceedings of the Royal Society of London*, **A 357**, 403-22.

[3] Burnell, J. G., Graham-Eagle, J.G., Gray, B. F. And Wake, G. C. Determination of critical ambient temperatures for thermal ignition. *I.M.A. Journal of Applied Mathematics*, 42, 147-54.

[4] Dancer, E. N. (1986). On the number of positive solutions of weakly nonlinear elliptic equations when a parameter is large. *Proceedings of the London Mathematical Society*, 53, 429-52.

[5] Frank-Kameneskii, D. A. (1955). *Diffusion and heat-transfer in chemical kinetics*, Princeton University Press, N.J.

[6] Gray, P. and Lee, P. R. *Thermal explosion theory*, Vol 2. Oxidation and Combustion Reviews Vol 2, Elsevier, London.

[7] Gray, P. and Sherrington, M. E. (1977). *Self-heating chemical kinetics and spontaneously unstable systems*, Specialist Periodical Reports, Gas Kinetics and Energy Transfer Vol **2**, The Chemical Society, London.

[8] Gray, P. and Wake, G. C. (1988). On the determination of critical ambient temperatures and critical ignition temperatures for thermal ignition. *Combustion and Flame*, **17**, 101-4.

Patterns, Waves and Interfaces in Excitable Reaction-Diffusion Systems

Masayasu Mimura
Hiroshima University

1 Introduction

Nonlinear reaction-diffusion models have been used widely to describe various phenomena in neurobiology, biophysics, chemical physics, population genetics, mathematical ecology and in other fields. One of the most striking features is the formation of propagating and non-propagating patterns. Under some circumstances, the system develops into qualitatively different states so that internal layers of interfaces are observed. Such interfaces exhibit a variety of geometrical patterns such as rotating spiral patterns in the Belousov-Zhabotinski reagent(Winfree 1980), dendritic patterns in solidifications (Caginalp 1986), pigmentation patterns on shells (Meinhardt and Klinger 1985) and animal coat marking (Murray 1981).

In these models the most simple and suggestive system is described by the following two-component model:

(1.1) $$\begin{cases} u_t = d_1 \Delta u + f(u,v) \\ v_t = d_2 \Delta v + \delta g(u,v). \end{cases}$$

With time and space variables t and x, $u(t,x)$ and $v(t,x)$ are called the activator and its inhibitor in morphogenesis or the propagator and its controller in excitable media. d_1 and d_2 are the diffusion rates of u and v. δ is the ratio of the reaction rates. The nonlinearities of f and g are roughly classified into three types, as in Figure 1(a), 1(b) and 1(c). A prototype of the nonlinearities of f and g is

(1.2) $$\begin{cases} f(u,v) = u(1-u)(u-a) - v \\ g(u,v) = u - \gamma v - I \end{cases}$$

with constants $0 < a < 1, \gamma > 0$ and $I > 0$. For a suitable choice of γ and I in (1.2), we have the above three cases.

The space-clamped version of (1.1) reduces to

(1.3) $$\begin{cases} \frac{du}{dt} = f(u,v) \\ \frac{dv}{dt} = \delta g(u,v). \end{cases}$$

which is often called the Bonhoeffer Van der Pol equation. For the case drawn in Figure 1(a), the system has only one equilibrium state P, which is globally stable. In this case, a small disturbance from this state is rapidly damped but a large disturbance wanders far from the state but then eventually returns to the state. Because of this feature, P is called a rest state in an excitable medium. For the case in Figure 1(b) also, the system has only one equilibrium state Q. However, the stability of this state depends on δ. When δ is large, it is stable, but when δ becomes small, it is unstable and there appear periodic solutions through Hopf bifurcation. Moreover, even if the state Q is stable for equation (1.3), it is not necessarily stable for equation (1.1). In fact, it is unstable when d_1 is small compared with d_2. This phenomenon is known as Turing's diffusion driven instability in morphogenesis. On the other hand, for the case in Figure 1(c), the system has three equilibria P, Q and R. Besides the stable state P, we have another stable state R and one unstable state Q, which are respectively called an excited state and a separator. Such a situation is called a bistable medium.

In this article, we treat the case when the first component u reacts much faster than the second v, although u diffuses more slowly than v. More specifically, we introduce the new parameters ε, τ and D by

$$\varepsilon = \sqrt{d_1}, \tau = \delta/\sqrt{d_1} \text{ and } D = d_2/\delta.$$

and rewrite (1.1) as

(1.4) $$\begin{cases} \varepsilon\tau u_t = \varepsilon^2 \Delta u + f(u,v) \\ v_t = D\Delta v + g(u,v). \end{cases}$$

Here we assume that ε is sufficiently small and τ and D are of the order $O(1)$.

The assumptions on the nonlinearities of f and g are described as follows:

A-1 $f = 0$ is S-shaped and consists of three branches $u = h_-(v)$, $h_0(v)$ and $h_+(v)$ ($h_-(v) \leq h_0(v) \leq h_+(v)$), while $g = 0$ intersects either only one of the branches or each branch once. The signs of f and g are negative in the regions lying above the curves $f = 0$ and $g = 0$. We denote by v_{max} and v_{min} relative maximum and relative minimum of $f(u,v) = 0$ (see Figure 1).

A-2 $J(v) \equiv \int_{h_-(v)}^{h_+(v)} f(u,v)du$ has a unique zero at $v = v^* \in \Sigma \equiv (v_{min}, v_{max})$.

A-3 $f_u(h_\pm(v), v) < 0$ and $g(h_-(v), v) < 0 < g(h_+(v), v)$ and $\frac{d}{dv}g(h_\pm(v), v) < 0$ for $v \in \Sigma$.

A-4 $f_v(u,v) < 0$ for $u \in (h_-(v), h_+(v))$ and $v \in \Sigma$.

In order to understand why internal layers or interfaces appear in (1.4) when ε is sufficiently small, we consider a simple version of (1.4) when v is constant, say $v \equiv \beta$ in the whole plane $\mathbf{R}^2$. i.e.,

(1.5) $$\varepsilon\tau u_t = \varepsilon^2 \Delta u + f(u,\beta).$$

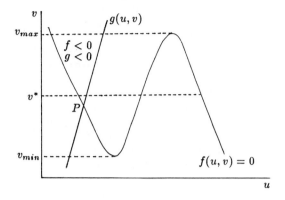
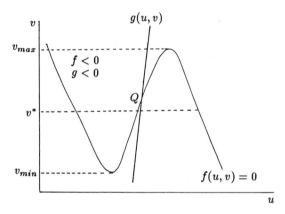
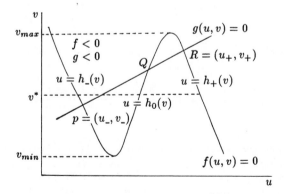

Figure 1. Functional forms of f and g

Here we assume β is such that $f(u,\beta)$ has three zeros $h_-(\beta)$, $h_0(\beta)$ and $h_+(\beta)$ and the equation is of bistable type. When ε is sufficiently small, the evolution of solutions of equation (1.5) would consist of two stages. The first stage is that even if the initial data $u(0,x)$ is smooth, the solution $u(t,x)$ tends, in a short time, to one of the stable equilibrium states $u \equiv h_+(\beta)$ in an x-region where $u(0,x) > h_0(\beta)$, while it tends to the other $u \equiv h_-(\beta)$ in a region where $u(0,x) < h_0(\beta)$ since the system is a bistable one. Thus, there appear internal layers with thickness $O(\varepsilon)$ separating two qualitatively different states. This is the birth of interfaces in the limit as $\varepsilon \downarrow 0$. The next stage is that such interfaces propagate. Suppose that the interface is described by a gentle curve $\Gamma(t)$ so that $\mathbf{R}^2 \backslash \Gamma(t) = \Omega_+(t) \cup \Omega_-(t)$ where $\Omega_+(t)$ and $\Omega_-(t)$ are disjoint regions and the relations $u \equiv h_\pm(\beta)$ hold in $\Omega_\pm(t)$. The motion of the interface is approximately described by the following evolution equation for $\Gamma(t)$:

$$(1.6) \qquad \tau V = \{-c_0(\beta) + \varepsilon K\} n$$

(for the derivation, see (Keener and Tyson 1986; Grindrod and Gomatam 1987) for instance). Here V is the normal velocity, K is the mean curvature at $\Gamma(t)$, n is the unit normal vector of $\Gamma(t)$ pointing from $\Omega_+(t)$ to $\Omega_-(t)$ and $c_0(\beta)$ is the velocity of the one dimensional travelling wave of the equation

$$w_t = w_{xx} + f(w,\beta)$$

with boundary conditions $w(\pm\infty) = h_\pm(\beta)$ for any fixed $\beta \in (v_{min}, v_{max})$. It is known that $c_0(\beta)$ is a strictly decreasing function of $\beta \in (v_{min}, v_{max})$ which satisfies

$$c_0(\beta) \gtreqless 0 \quad \text{if and only if} \quad J(\beta) \gtreqless 0$$

(Fife and McLeod 1977). If $f(u,v)$ takes the form (1.2), $c_0(\beta)$ is explicitly represented as

$$c_0(\beta) = \frac{h_+(\beta) - 2h_0(\beta) + h_-(\beta)}{\sqrt{2}}.$$

Several authors have investigated the curvature effect on the motion of interfaces (Gage and Hamilton 1986; Grindrod 1988; Rubinstein et al. 1989) and the references therein. More recently, Evans and Spruck (1989) and Chen(1989) have studied the global existence and asymptotic behaviour of $\Gamma(t)$ of the initial value problem to (1.6).

We should note that when $c_0(\beta) = 0$, equation (1.6) does not describe the dynamics of interfaces in one dimension. Another approximating equation for interfaces has to be proposed in place of (1.6). We will not discuss this case. The reader should refer to (Kawasaki and Ohta 1982) and (Carr and Pego 1988).

In Section 2, we derive the equation of motion for an interface from the system (1.4). This is a free boundary problem with unknowns v and the interface (see (2.3)). In considering the dynamics of internal layers and interfaces, the following question arises: what is the relation between the equation for the interface and reaction-diffusion systems (1.4) in which internal layers appear with the thickness $O(\varepsilon)$, in the limit as $\varepsilon \downarrow 0$?

As the first step in answering this question, we consider two special types of solutions: (1) travelling wave solutions in a bistable medium in Section 3 and (2) stationary pulse solutions in Section 4. We show that for these solutions, the equation for an interface is a nice approximation to the behaviour of reaction-diffusion systems in the limit as $\varepsilon \downarrow 0$. This is made precise in Theorems 3.1 and 4.1.

2 The equation of motion for an interface

Consider the system of equations (1.4) in $\mathbf{R}^2$. Keeping the scalar equation (1.2) in mind, one would expect that the evolution of solutions of (1.4) consists of two stages: the first stage is the occurrence of internal layers (or interfaces in the limit as $\varepsilon \downarrow 0$). In the region away from an interface, we may formally put $\varepsilon = 0$ in (1.4) so that it becomes

(2.1)
$$\begin{cases} f(u,v) &= 0 \\ v_t &= D\Delta v + g_\pm(v) \end{cases}$$

where $g_\pm(v) = f(h_\pm(v), v)$. The functions $u = h_\pm(v)$ correspond to the two branches of $f(u,v) = 0$ as shown in Figure 1. Thus, the whole region $\mathbf{R}^2$ is decomposed into two different regions Ω_+ and Ω_- in which $u = h_+(v)$ and $u = h_-(v)$ hold, respectively. The boundary between Ω_+ and Ω_- is an interface, say $\Gamma(t)$. We next consider the motion of the interface as the second stage. As for the scalar version, we have

(2.2)
$$\tau V = \{-c_0(\beta) + \varepsilon K\}n.$$

Here β is the value of v on the interface. We thus have the equation of motion for an interface

(2.3)
$$\begin{cases} v_t &= D\Delta v + g_\pm(v) \quad (t,x) \in \Omega_\pm \\ \tau V &= -\{c_0(\beta) + \varepsilon K\}n \quad \text{on } \Gamma(t) \end{cases}$$

with suitable continuity conditions for v on the interface (Fife (1985), (1988)). In the system (2.3), the unknowns are the bulk variable v and the interface Γ, i.e., (2.3) is a free boundary problem.

For the one dimensional case, Hilhorst et al. (1989) have recently proved the global existence of a solution $v(t,x)$ and more recently Chen (1989) has shown that a solution of (2.3) exists locally for higher dimensional cases.

For the dynamics of interfaces, there is a great difference between the scalar case (1.6) and the system case (2.3). If an initial interface is convex, it is *still* convex for any time under (1.6) because of the curvature effect (Chen 1989), while it possibly becomes non-convex under (2.3) if the interaction effect of v on the interface overcomes the curvature effect.(Figure 2). For this reason, the analysis of (2.3) seems to be considerably harder.

Figure 2. Dynamics of a slightly deformed spherical interface to (2.3) for (1.2) with $a = 0.25$, $\gamma = 6$, $\tau = 0.7$.

We thus have introduced two different systems. One, equation (1.4), is the so-called reaction diffusion system with a small parameter $\varepsilon > 0$ and the other (2.3) is the interface equation associated with (1.4) in the limit as $\varepsilon \downarrow 0$. The relation between (1.4) and (2.3) has up to now been studied by heuristic and asymptotic analysis but no rigorous understanding has been achieved. The purpose of this article is to discuss this relation for particular solutions such as travelling front solutions and stationary pulse solutions.

3 Travelling front solutions in bistable media

In this section, we assume f and g are bistable nonlinearities as in Figure 1(c) and consider one dimensional travelling front solutions which correspond to the propagation of transition from one stable state P to the other R. These waves can be represented as $(u,v)(z)$ with $z = x + ct$ where the velocity of the wave equals c. Thus (1.4) can be rewritten as

$$(3.1) \quad \begin{cases} \varepsilon^2 u_{zz} - \varepsilon c \tau u_z + f(u,v) = 0 \\ v_{zz} - c v_z + g(u,v) = 0. \end{cases}, z \in \mathbf{R}$$

where we may take $D = 1$ without loss of generality. The boundary conditions for (u,v) are

$$(3.2) \quad \begin{cases} u(\pm\infty) = u_\pm \\ v(\pm\infty) = v_\pm. \end{cases}$$

We expect that the wave front possesses a single internal layer with the thickness $O(\varepsilon)$, which becomes an interface when $\varepsilon \downarrow 0$.

On the other hand, the problem for the travelling(propagating) interface is also formulated by making use of the moving coordinate $z = x + ct$. When the location of the interface is taken as $z = 0$, (2.3) becomes

$$(3.3) \quad \begin{cases} v_{zz}^\pm - c v_z^\pm + g_\pm(v^\pm) = 0, \quad z \in \mathbf{R}_\pm \\ \tau c = c_0(\beta). \end{cases}$$

The boundary conditions are

$$(3.4) \quad \begin{cases} v^-(0) = v^+(0) = \beta \\ v_z^-(0) = v_z^+(0) \\ \lim_{z \to \pm\infty} v^\pm(z) = v_\pm, \end{cases}$$

where β is the value of v on the interface, which will be determined later. The problem is to find c and β such that $v^\pm$ satisfy (3.3) and (3.4). We call such c the velocity of the interface. We first note the following lemmas:

Lemma 3.1 (Ikeda et al. 1989) *For any fixed $c \in \mathbf{R}$ and $\beta \in (v_-, v_+)$, there exist unique strictly monotone increasing solutions $v_0^\pm(z; c, \beta)$ of (3.3), which satisfy*

$$(3.5) \quad \frac{\partial}{\partial c}[\frac{d}{dz} v_0^-(0; c, \beta) - \frac{d}{dz} v_0^+(0; c, \beta)] > 0$$

and

$$(3.6) \quad \frac{\partial}{\partial \beta}[\frac{d}{dz} v_0^-(0; c, \beta) - \frac{d}{dz} v_0^+(0; c, \beta)] > 0.$$

Lemma 3.2 (Ikeda et al. 1989) *There exists a unique strictly monotone decreasing function $\beta = \beta_0(c)$ for $c \in \mathbf{R}$ satisfying*

$$\frac{d}{dz}v_0^-(0;c,\beta_0(c)) - \frac{d}{dz}v_0^+(0;c,\beta_0(c)) = 0,$$

and it converges to $v_\pm$ as $c \to \mp\infty$.

Thus, we easily find that the problem (3.3), (3.4) is equivalent to

(3.7) $\qquad\qquad \beta = \beta_0(c) \quad \text{and} \quad c = c_0(\beta)/\tau.$

By solving (3.7), we arrive at the following theorem:

Theorem 3.3 *Consider the problem (3.3), (3.4). There is at least one solution for any $\tau > 0$. In particular, when $v^* \in (v_-, v_+)$, there are three solutions for small τ, while only one for large τ. On the other hand, when $v^* \in (v_{min}, v_{max})\backslash(v_-, v_+)$, there is only one solution for small or large τ.*

This result implies the occurrence of bifurcation phenomena of travelling interfaces when τ is varied over an appropriate range of values. In fact, if we use the kinetics (1.2), Theorem 3.3 can be more clearly understood. We fix arbitrarily a to satisfy $0 < a < \frac{1}{2}$ and define

$$\gamma_0 = 9/\{(2-a-\sqrt{a^2-a+1})(1-2a+\sqrt{a^2-a+1})\},$$

$$\gamma_1 = 9/\{(2-a)(1-2a)\},$$

$$\gamma_2 = 9/(1+a+\sqrt{3(a^2-a+1)})/\{(1+a)(2-a)(1-2a)\}.$$

It is easy to see that the situations where $v^* \in (v_-, v_+)$ and $v^* \in (v_{min}, v_{max})\backslash(v_-, v_+)$ correspond to $\gamma_0 < \gamma < \gamma_2$ and $\gamma_2 < \gamma$, respectively (Figure 3).

The global bifurcation pictures with respect to τ are drawn in Figure 4 for some values of γ. We note that the limit point in Figures 4(a) and 4(c) corresponds to the point at which the two curves in (3.7) intersect tangentially. When $\gamma = \gamma_1$ so that the kinetics (1.2) possesses odd symmetry, there exists an odd symmetric standing wave solution for any τ, and a pitchfork bifurcation occurs at some value of τ (Figure 4(b)). When γ is slightly different from γ_1, this symmetric structure is deformed into the well known structures of imperfect bifurcation theory. When γ is greater than γ_2, the value of τ at which the limit point appears tends to zero so that there is only one solution branch for any τ (Figure 4(d)).

We now consider the problem (3.1), (3.2) for small $\varepsilon > 0$. Using the functions $v_0^\pm(z;c,\beta)$ obtained in Lemma 3.1, we define $u_0^\pm(z;c,\beta)$ by

$$u_0^\pm(z;c,\beta) = h_\pm(v_0^\pm(z;c,\beta)), z \in \mathbf{R}_\pm,$$

which is discontinuous at $z = 0$. We call $(u_0^\pm(z,c,\beta), v_0^\pm(z,c,\beta))$ a singular limit travelling front solution. Since it does not satisfy (3.1) in a neighbourhood of $z = 0$,

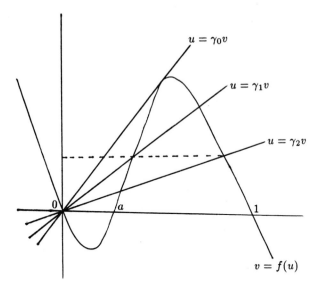

Figure 3. Critical values of γ

we use new functions $w^\pm$ such that $(u_0^\pm + w^\pm, v_0^\pm)$ are approximating solutions in the semi-infinite intervals $\mathbf{R}_\pm$. We introduce a stretched variable $\xi = z/\varepsilon$ in a neighbourhood of $z = 0$, and substitute $(u_0^\pm + w^\pm, v_0^\pm)$ into (3.1). Then, putting $\varepsilon = 0$, we obtain the equations for $w^\pm$

(3.8)
$$\begin{cases} w_{\xi\xi}^\pm - c\tau w_\xi^\pm + f(h_\pm(\beta) + w^\pm, \beta) = 0, & \xi \in \mathbf{R}_\pm \\ w^\pm(0) = \alpha - h_\pm(\beta) \\ w^\pm(\pm\infty) = 0. \end{cases}$$

Here we normalized $u(z)$ by the condition $u(0) = \alpha$ for suitable fixed α.

Lemma 3.4 (Ikeda et al. 1989) *For any fixed $\beta^* \in (v_-, v_+)$, let $c^*(\tau) = c_0(\beta^*)/\tau$. Then there exists $\delta_0 > 0$ such that for any fixed $(c, \beta) \in \Lambda_{\delta_0} = \{(c, \beta) \mid |c - c^*(\tau)| + |\beta - \beta^*| \leq \delta_0\}$, (3.8) have unique strictly monotone increasing solutions $w^\pm(\xi; \tau; c, \beta)$.*

Let $(c^*(\tau), \beta^*(\tau))$ be an arbitrary intersection of the curves in (3.7) and define $I(\tau)$ by

$$I(\tau) = \frac{d}{dc}\beta_0(c^*(\tau)) - \tau(\frac{d}{d\beta}c_0(\beta^*(\tau)))^{-1},$$

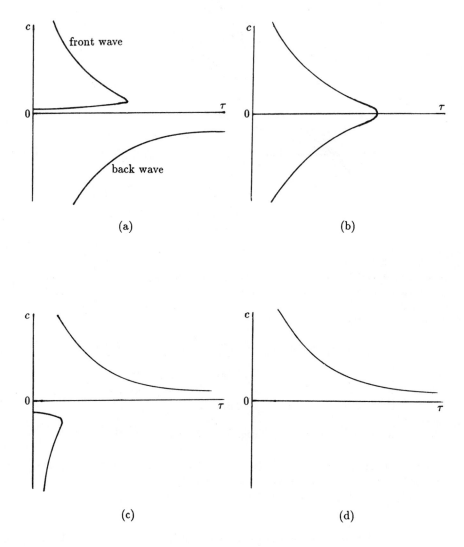

Figure 4. The velocity of travelling interface versus τ for different values of γ. (a) $\gamma_0 < \gamma < \gamma_1$. (b) $\gamma = \gamma_1$. (c) $\gamma_1 < \gamma < \gamma_2$. (d) $\gamma_2 \leq \gamma$.

which is the difference of the gradients of the two curves in (3.7) at $(c, \beta) = (c^*(\tau), \beta^*(\tau))$. If these curves intersect tangentially, then $I(\tau) = 0$, while if the curves intersect transversally, then $I(\tau) \neq 0$.

We also define

$$u_0(z; \varepsilon, \tau) = \begin{cases} u_0^-(z; c^*(\tau), \beta^*(\tau)) + w^-(z/\varepsilon; \tau; c^*(\tau), \beta^*(\tau)), & z \in \mathbf{R}_- \\ u_0^+(z; c^*(\tau), \beta^*(\tau)) + w^+(z/\varepsilon; \tau; c^*(\tau), \beta^*(\tau)), & z \in \mathbf{R}_+ \end{cases}$$

and

$$v_0(z; \varepsilon, \tau) = \begin{cases} v^-(z; c^*(\tau), \beta^*(\tau)), & z \in \mathbf{R}_- \\ v^+(z; c^*(\tau), \beta^*(\tau)), & z \in \mathbf{R}_+ \end{cases}$$

Theorem 3.5 *Suppose that (A-1)-(A-4) hold and fix τ arbitrarily such that $(c^*(\tau), \beta^*(\tau))$ satisfies $I(\tau) = 0$. Then there is $\varepsilon_0 > 0$ such that for any $\varepsilon \in (0, \varepsilon_0)$, (3.1) ,(3.2) has a travelling wave solution $(U(z; \varepsilon; \tau), V(z; \varepsilon; \tau))$ with velocity $c(\varepsilon; \tau)$. Moreover, as $\varepsilon \downarrow 0$*

$$\|U(\cdot; \varepsilon; \tau) - u_0(\cdot; \varepsilon; \tau)\|_{X^1_{\rho, \varepsilon}(\mathbf{R})} + \|V(\cdot; \varepsilon; \tau) - v_0(\cdot; \varepsilon; \tau)\|_{X^1_{\rho, 1}(\mathbf{R})} \to 0$$

for some $\rho > 0$ where

$$\|u\|_{X^p_{\rho,\sigma}(\mathbf{R})} = \sum_{i=0}^{p} \sup_{x \in \mathbf{R}} |e^{\rho|x|} \left(\sigma \frac{d}{dx}\right)^i u(x)|$$

and

$$c(\varepsilon, \tau) \to c^*(\tau).$$

Consequently, the relation is revealed between the problems for $\varepsilon > 0$ and $\varepsilon \downarrow 0$. When τ is varied in $(0, +\infty)$, we can draw the global picture of travelling waves which is almost the same as that of the limiting case $\varepsilon \downarrow 0$ except for the bifurcation point or the limit point. The interesting point is the coexistence of a front wave and a back wave under some circumstances, as in Figure 4. This is an essential difference from the scalar version which possesses only one travelling wave.

We now discuss the stability of these solutions. The evolution equation associated with (3.1) is

(3.9) $$\begin{cases} u_t = \varepsilon^2 u_{zz} - \varepsilon c \tau u_z + f(u, v) \\ v_t = v_{zz} - c v_z + g(u, v) \end{cases} \quad t > 0, z \in \mathbf{R}$$

with the boundary conditions

(3.10)
$$\begin{cases} u(t, \pm\infty) = u_\pm \\ v(t, \pm\infty) = v_\pm. \end{cases}$$

Theorem 3.6 (Nishiura et al. 1989b.) *Let $(c^*(\tau), \beta^*(\tau))$ be an intersection point of (3.7). Then the stationary solution $(U(z;\varepsilon;\tau), V(z;\varepsilon;\tau))$ of (3.9), (3.10) with velocity $c(\varepsilon;\tau)$ is asymptotically stable if $I(\tau) > 0$, while it is unstable if $I(\tau) < 0$, where $\lim_{\varepsilon \downarrow 0} c(\varepsilon;\tau) = c^*(\tau)$.*

This result says that information on the sign of $I(\tau)$ obtained from the limiting problem ($\varepsilon \downarrow 0$), plays an important role in determining the stability as well as the existence of travelling wave solutions of the reaction-diffusion problem ($\varepsilon > 0$).

This assertion is close to the spirit of (Evans 1972, 1975) where the stability of a nerve impulse is discussed in terms of the intersection of stable and unstable manifolds (see also (Ikeda 1989)). One unsolved problem is the case when $I(\tau) = 0$. This is a problem for further study.

We are next interested in a planar travelling front solution in $\mathbf{R}^2$ of the form

$$(u(x,y,t), v(x,y,t)) = (U(z), V(z))$$

with $z = (x,y) \cdot \nu + ct$ for some unit vector $\nu \in \mathbf{R}^2$. The boundary conditions are

$$\lim_{z \to \pm\infty} (U(z), V(z)) = (u_\pm, v_\pm).$$

Thus (U, V) is the 1D travelling front solution constructed in Theorem 3.5 and then Theorem 3.6 gives almost complete information on the stability properties in one dimension. A natural question arises whether this planar travelling front (or interface) is stable or not in higher dimensions. We have recently found that the stability essentially depends on the parameter τ. This will be described elsewhere.

4 Standing pulse solution

In this section, we assume the nonlinearities f and g are as in Figure 1(b), i.e., there is only one constant equilibrium state $P = (u_-, v_-)$. In addition, we assume

$$\int_{h_-(v_-)}^{h_+(v_-)} f(u, v_-) du > 0,$$

so that the equilibrium state P is locally stable. When we take the kinetics of (1.2), the above condition is satisfied for $0 < a < \frac{1}{2}$, $0 < \gamma < \max_{u>0}(u-a)(1-u)$ and $I = 0$. In this situation, there occurs a localization of u when the rate of diffusion of v

is large relative to that of u. This process is heuristically understood as follows: first a local perturbation in u forms into a large peak and expands, as an activator, but its propagation may be stopped due to the relatively fast diffusion of the inhibitor. This mechanism is well known as "lateral inhibition". Mathematically, it implies that there exists a (hopefully stable) stationary pulse solution when the ratio of the diffusion rates of u and v is very small. This motivates us to study the following 1D stationary problem:

(4.1) $$\begin{cases} 0 = \varepsilon^2 u_{xx} + f(u,v) \\ 0 = v_{xx} + g(u,v) \end{cases} \quad -\infty < x < \infty$$

with the boundary conditions

(4.2) $$\lim_{|x| \to \infty} (u,v) = (u_-, v_-).$$

Here we may assume $D = 1$. If we restrict to symmetric solutions with respect to $x = 0$, the problem (4.1), (4.2) is written as

(4.3) $$\begin{cases} 0 = \varepsilon^2 u_{xx} + f(u,v) \\ 0 = v_{xx} + g(u,v) \end{cases} \quad 0 < x < \infty$$

with the boundary conditions

(4.4) $$(u_x, v_x)(0) = 0, \quad \lim_{x \to \infty}(u,v) = (u_-, v_-).$$

On the other hand, the limiting problem as $\varepsilon \downarrow 0$ corresponding to (4.3), (4.4) is

(4.5) $$\begin{cases} v^{\pm}_{xx} + g^{\pm}(v^{\pm}) = 0 \\ 0 = \tau V = c_0(\beta), \end{cases} \quad z \in I_{\pm}$$

where $I_+ = (0, \xi)$ and $I_- = (\xi, \infty)$ with an unknown value ξ which will be determined later. The boundary conditions are

(4.6) $$\begin{cases} v_x^+(0) = 0, & \lim_{x \to \infty} v^-(x) = 0 \\ v^+(\xi) = v^-(\xi), & v_x^+(\xi) = v_x^-(\xi). \end{cases}$$

The problem in the limit $\varepsilon \downarrow 0$ is to find β and ξ such that $v^{\pm}(x)$ satisfy (4.5), (4.6). This problem is easily solved as follows: by (A.2), we obtain v^* to satisfy $c_0(v^*) = 0$ and then solve (4.5), (4.6) so as to find ξ_*, which is a stationary interface, such that $v_0^{\pm}(x)$ are solutions of (4.5), (4.6).

Before considering the problem (4.3), (4.4), we define $(\bar{u}_0(x), \bar{v}_0(x))$ by

$$\bar{u}_0(x) = \begin{cases} h_+(v_0^+(x)), & x \in I_+ \\ h_-(v_0^-(x)), & x \in I_- \end{cases}$$

and

$$\bar{v}_0(x) = \begin{cases} v_0^+(x), & x \in I_+ \\ v_0^-(x), & x \in I_-. \end{cases}$$

Using shooting arguments (Ermentrout et al. 1984) or singular perturbation techniques, we can find at least two different (nonconstant) solutions of (4.1), (4.2) for small ε. More precisely, we have

Theorem 4.1 *There is $\varepsilon_0 > 0$ such that (4.1), (4.2) has two pulse solutions $(\bar{u}_\varepsilon(x), \bar{v}_\varepsilon(x))$ and $(\underline{u}_\varepsilon(x), \underline{v}_\varepsilon(x))$ which are symmetric with respect to $x = 0$ for each $\varepsilon \in (0, \varepsilon_0)$. Moreover,*

$$\begin{cases} \lim_{\varepsilon \downarrow 0} \bar{u}_\varepsilon(x) = \bar{u}_0(x), & x \in \mathbf{R} \setminus (\xi_* - \bar{\delta}, \xi_* + \bar{\delta}) \cup (-\xi_* - \bar{\delta}, -\xi_* + \bar{\delta}) \\ \lim_{\varepsilon \downarrow 0} \bar{v}_\varepsilon(x) = \bar{v}_0(x), & x \in \mathbf{R} \end{cases}$$

$$\begin{cases} \lim_{\varepsilon \downarrow 0} \underline{u}_\varepsilon(x) = 0, & x \in \mathbf{R} \setminus (-\underline{\delta}, \underline{\delta}), \\ \lim_{\varepsilon \downarrow 0} \underline{v}_\varepsilon(x) = 0, & x \in \mathbf{R} \end{cases}$$

where $\bar{\delta}$ and $\underline{\delta}$ are arbitrary positive constants (see Figure 5).

We now address the following questions: (1) Are there any pulse solutions except for the above two solutions for small ε? (2) Are there any pulse solutions for larger ε? Unfortunately, these have not yet been answered except for the case when (f, g) are piecewise-linear

(4.7)
$$\begin{cases} f(u, v) = -1 + H(u - a) - v \\ g(u, v) = u - \gamma v \end{cases}$$

with constants a and γ satisfying $0 < a < \frac{1}{2}$ and $0 \leq \gamma < 2a/(1-2a)$. For this case, the solution branch of pulse solutions can be drawn as in Figure 6, i.e., there is the critical value ε_c such that there are two different pulse solutions for $0 < \varepsilon < \varepsilon_c$, where the upper and lower branches correspond to $(\bar{u}_\varepsilon(x), \bar{v}_\varepsilon(x))$ and $(\underline{u}_\varepsilon(x), \underline{v}_\varepsilon(x))$, respectively, while there is no solution for $\varepsilon_c < \varepsilon$.

The stability of these solutions can be investigated by using the singular limit eigenvalue problem (SLEP) method originally developed by Nishiura and Fujii (1987).

Theorem 4.2 *Consider (4.1), (4.2) with sufficiently small ε. There is a critical value τ_c such that the large solution $(\bar{u}_\varepsilon(x), \bar{v}_\varepsilon(x))$ (given in Theorem 4.1) is asymptotically stable for $\tau_c < \tau$ and unstable for $0 < \tau < \tau_c$. On the other hand, the small solution $(\underline{u}_\varepsilon(x), \underline{v}_\varepsilon(x))$ is unstable for any $\tau > 0$.*

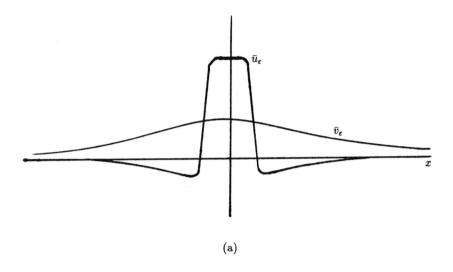

(a)

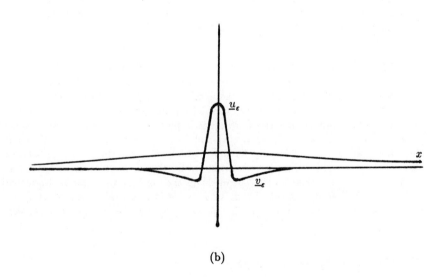

(b)

Figure 5. (a) $(\bar{u}_\varepsilon, \bar{v}_\varepsilon)$-solution. (b) $(\underline{u}_\varepsilon, \underline{v}_\varepsilon)$-solution.

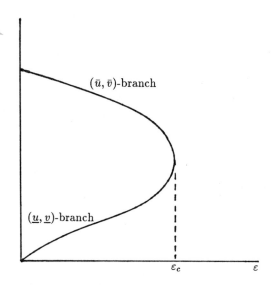

Figure 6. Solution branch of symmetric pulse solutions with respect to ε.

Remark 4.1 When τ is decreasing near $\tau = \tau_c$, the destabilization occurs through Hopf bifurcation, and for fixed τ smaller than τ_c, there appears the breather solution (Koga and Kuramoto 1980 ; Nishiura and Mimura 1989a).

It is quite interesting to consider stationary pulse solutions in higher dimensions. Recently Ohta et al. (1989) have studied the stability of several types of stationary pulse solutions in $\mathbf{R}^2$ and $\mathbf{R}^3$ when the nonlinearities f and g are given by (4.7). An interesting result is the stability of radially symmetric solutions. The solution branches of 1D- and 2D-spherically symmetric pulse solutions with respect to the parameter a are shown in Figures 7(a) and (b), respectively. As was stated in the above, when τ is arbitrarily fixed to satisfy $\tau > \tau_c$, the 1D large solution $(\bar{u}_\varepsilon(x), \bar{v}_\varepsilon(x))$ is stable. However, the situation for the 2D large solution is different as the stability of the large solution depends upon a. In fact, there exists the critical value a_c such that it is stable for $a_c < a$, while it is unstable and a symmetry breaking solution occurs for $a < a_c$. For general f and g, the study of this problem is in progress.

5 Concluding remarks

For travelling front wave solutions and stationary pulse solutions, we can say that the equation for interfaces (2.3) is the limiting system for the reaction-diffusion

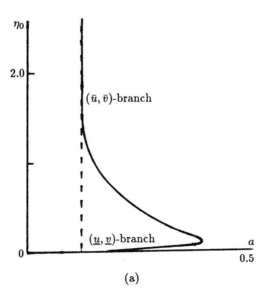

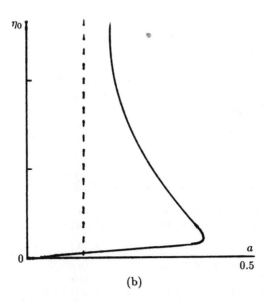

Figure 7. Solution branch of symmetric solutions with respect to a. $\varepsilon = 0.005$, $\gamma = 1/3$. η_0 is the distance between $x = 0$ and the location of the interface.

system (1.4). However, we have not been able to say anything on the general initial value problems to (1.4) and (2.3) in higher dimensions. This is a problem which we shall study in future.

6 Acknowledgement

I have benefitted from discussions with H. Fujii, Y. Nishiura, H. Ikeda, T. Tsujikawa, R. Kobayashi and T. Ohta.

References

[1] Caginalp, G. (1986). An analysis of a phase field model of a free boundary. *Archive for Rational Mechanics and Analysis*, **92**, 205-45.

[2] Carr, J. and Pego, R. L. (1989). Metastable patterns in solutions of $u_t = \varepsilon^2 u_{xx} - f(u)$. *Communications on Pure and Applied Mathematics*, **42**, 523-76.

[3] Chen, X.-Y. (1989). Dynamics of interfaces in reaction diffusion systems. To appear in *Hiroshima Mathematical Journal*.

[4] Chen, Y. G., Giga, Y. and Goto, S. (1989). Uniqueness and existence of viscosity solutions of generalized mean curvature flow equations. *Proceedings of Japan Academy*, **65**, 207-210.

[5] Dockery, J. D. and Keener, J. D. (1989). Diffusive effects on dispersion in excitable media. *SIAM Journal on Applied Mathematics*, **49**, 539-66.

[6] Ermentrout, G. B., Hastings, S. P. and Troy, W. C. (1984). Large amplitude stationary waves in an excitable lateral-inhibitory medium. *SIAM Journal on Applied Mathematics*, **44**, 1133-49.

[7] Evans, L. C. and Spruck, J. (1989). Motion of level sets by mean curvature. To appear.

[8] Evans, J. W. (1972). Nerve axon equations, III: Stability of the nerve impulse. *Indiana University Mathematics Journal*, **21**, 577-93.

[9] Evans, J. W. (1975). Nerve axon equations, IV: The stable and the unstable impulse. *Indiana University Mathematics Journal*, **24**, 116-90.

[10] Fife, P. C. (1985). Understanding the patterns in the BZ reagent. *Journal of Statistical Physics*, **39**, 687-703.

[11] Fife, P. C. (1988). *Dynamics of internal layers and diffusive interfaces*. CBMS-NSF Regional Conference Series in Applied Mathematics, **53**, SIAM, Philadelphia.

[12] Fife, P. C. and McLeod, J. B. (1977). The approach of solutions of nonlinear diffusion equations to travelling wave solutions. *Archive for Rational Mechanics and Analysis*, **63**, 335-61.

[13] Gage, M. and Hamilton, R. S. (1986). The heat equation shrinking convex plane curves. *Journal of Differential Geometry*, **23**, 69-96.

[14] Grindrod, P. (1988). Geometrical theory for wave propagation in reaction-diffusion models. To appear in *Mathematical Biosciences*.

[15] Grindrod, P. and Gomatam, J. (1987). The geometry and motion of reaction-diffusion waves on closed two dimensional manifolds. *Journal of Mathematical Biology*, **25**, 597-610.

[16] Hilhorst, D., Nishiura, Y. and Mimura, M. (1989). A free boundary problem arising in reaction-diffusion systems. To appear.

[17] Ikeda, H. (1989). Singular perturbation approach to stability properties of travelling wave solutions of reaction-diffusion systems. *Hiroshima Mathematical Journal*, **19**, 587-630.

[18] Ikeda, H., Mimura, M. and Nishiura, Y. (1989). Global bifurcation phenomena of travelling wave solutions for some bistable reaction-diffusion systems. *Nonlinear Analysis*, **13**, 507-26.

[19] Koga, S., and Kuramoto, Y. (1980). Localized patterns in reaction-diffusion systems. *Progress of Theoretical Physics*, **63**, 106-21.

[20] Kawasaki, K. and Otha, T. (1982). Kink dynamics in one-dimensional nonlinear dynamics. *Physica A*, **116**, 573-93.

[21] Keener, J. R. and Tyson, J. J. (1986). Spiral waves in the Belousov-Zhabotinski reaction. *Physica D*, **21**, 307-24.

[22] Meinhardt, H. and Klinger, M. (1985). Pattern formation by coupled oscillations: The pigmentation pattern on the shells of molluscs. *Lecture Notes in Biomathematics*, **71**, 184-98.

[23] Murray, J. D. (1981). A pre-pattern formation for animal coat markings. *Journal of Mathematical Biology*, **88**, 161-99.

[24] Nishiura, Y. and Fujii, H. (1987). Stability of singularly perturbed solutions to systems of reaction-diffusion equations. *SIAM Journal on Mathematical Analysis*, **18**, 1726-70.

[25] Nishiura, Y. and Mimura, M. (1989a). Layer oscillations in reaction-diffusion systems. *SIAM Journal on Applied Mathematics*, **49**, 515-39.

[26] Nishiura, Y., Mimura, M., Ikeda, H. and Fujii, H. (1989b). Singular limit analysis of stability of travelling wave solutions in bistable reaction-diffusion systems. To appear in *SIAM Journal on Mathematical Analysis*.

[27] Ohta, T., Mimura, M. and Kobayashi, R. (1989). Higher-dimensional localized pattern in excitable media. *Physica D*, **34**, 115-44.

[28] Rubinstein, J., Sternberg, P. and Keller, J. B. (1989). Fast reaction, slow diffusion and curve shortening. *SIAM Journal on Applied Mathematics*, **49**, 116-33.

[29] Winfree, A. T. (1980). *The geometry of biological time*. Springer, Berlin.

Reduced equivariant Conley index and applications

J. Smoller [1]

A. Wasserman [2]

University of Michigan

1 Introduction

In this article we sidestep the difficulties of defining and computing the Conley index of an isolated invariant set in an infinite-dimensional setting, by constructing a global Lyapunov-Schmidt reduction to a finite-dimensional space. This enables us to use the finite-dimensional version of the Conley Index Theory. In particular, we enrich this invariant by using the group structure, present from symmetry considerations, to obtain an equivariant Conley index. We use this equivariant Conley index to prove a general bifurcation theorem which is then applied to the symmetry-breaking problem for semilinear elliptic equations. (See the papers of Amann and Zehnder(1980), Chow and Lauterbach(1985), and Kielhofer(1988) for other ways of reducing to finite-dimensional problems in analogous situations.)

The framework can be described as follows. Let B_0 and B_1 be Banach spaces, and let H be a Hilbert space, where $B_1 \subset B_0 \subset H$ and the inclusions are all assumed to be continuous. Let $I = [\lambda_1, \lambda_2]$ be an interval in $\mathbf{R}$ and suppose that $M : B_1 \times I \to B_0$ is a smooth, gradient operator and that there exists a smooth curve $\{u_\lambda : \lambda \in I\} \subset B_1$ satisfying

(1.1) $$M(u_\lambda, \lambda) = 0, \ \lambda \in I.$$

For $i = 1, 2$, denote by P_i the (closed) vector space spanned by

$$\{v \in B_1 : d_u M(u_{\lambda_i}, \lambda_i)v = \mu v, \ \mu \geq 0\};$$

P_i is called the *peigenspace* of $(u_{\lambda_i}, \lambda_i)$. We first assume that the following conditions hold:

(1.2) $$dM_i \equiv d_u M(u_{\lambda_i}, \lambda_i) \text{ is nonsingular for } i = 1, 2,$$

and

(1.3) $$\dim P_1 \neq \dim P_2.$$

---
[1] Research supported in part by the NSF, under Grant No. MCS 830123.
[2] Research supported in part by the ONR, under Grant No. N0014-88-K0082.

The question we ask is : Does there exist λ, $\lambda_1 < \lambda < \lambda_2$ such that (u_λ, λ) is a bifurcation point for (1.1) ? Notice that in the finite-dimensional case (i.e., B_1 and B_0 have finite dimension), the Conley indices satisfy $h(u_{\lambda_1}) \neq h(u_{\lambda_2})$ so Conley's Continuation Theorem, see (Conley 1978; Smoller 1983), implies that our question has an affirmative answer.

More generally, suppose that G is a compact Lie group acting on H with B_0 and B_1 invariant under G, and that M is equivariant with respect to G in the sense that

(1.4) $$M(gu, \lambda) = gM(u, \lambda), \forall g \in G, u \in B_1, \lambda \in I.$$

Then the set of solutions of $M = 0$ also admits a G-action. If $\{(u_\lambda, \lambda) : \lambda \in I\}$ is a smooth curve of *invariant* solutions[3] under G, (in the sense that $gu_\lambda = u_\lambda$ $\forall g \in G, \lambda \in I$), we pose the question as to whether the symmetry breaks along this curve, i.e., under what conditions does there exist bifurcation from this curve of invariant solutions to noninvariant solutions?

In order to describe our approach to this problem, we first note that (1.4) implies that for all $g \in G$, $v \in B_1$, and each λ,

$$d_u M(gu, \lambda) gv = g d_u M(u, \lambda) v.$$

If now V denotes the eigenspace of $d_u M(u_\lambda, \lambda)$ corresponding to a given eigenvalue μ, then if $w \in V$, we have that $gw \in V$ for all $g \in G$. Indeed,

$$d_u M(u_\lambda, \lambda) w = \mu w \text{ implies } g d_u M(u_\lambda, \lambda) w = \mu gw,$$

so that

$$d_u M(u_\lambda, \lambda) gw = d_u M(gu_\lambda, \lambda) gw = \mu gw,$$

since $gu_\lambda = u_\lambda$, so $gw \in V$. Thus if T_g is the operator on V defined by $T_g v = gv$, then the mapping $g \to T_g$ defines a *representation* of G. In the language of representation theory, we say that V is a *representation* of G. It follows that for each fixed λ, the peigenspace of $d_u M(u_\lambda, \lambda)$ defines a representation of G. This motivates us to replace condition (1.3) by the statement

(1.3)$_G$ P_1 is not isomorphic to P_2 as a representation of G.

Now we ask this question: Suppose that (1.1) and (1.3)$_G$ hold; does this imply that bifurcation occurs? Observe that condition (1.3)$_G$ is considerably weaker than (1.3). Indeed, it is possible that (1.3)$_G$ holds even though P_1 and P_2 are isomorphic vector spaces.

We will consider such questions from the Conley index point of view. This requires that we extend these ideas to the equivariant case, and we discuss this in §3. Moreover, the hypotheses (1.2) and (1.3), or (1.3)$_G$ do not directly imply, in the infinite-dimensional case, that there is a change in Conley index at λ_1 and λ_2. Thus in §2 we describe a new, global, equivariant Lyapunov-Schmidt reduction,

---
[3] sometimes called symmetric solutions

which enables us to prove a general bifurcation theorem which answers affirmatively the above questions. In §4 we shall show how our results apply to the symmetry-breaking problem for semi-linear elliptic equations defined on n-balls. In the final section we mention several related results, and we also show how our results in §2 be used to clarify some subtle points in (Smoller 1983).

This article is an explanatory survey of some of the results in (Smoller and Wasserman 1990a,b). Complete proofs of our statements given here can be found in these papers. We would like to thank A. Blass for showing us the proof of the result given in the Appendix.

2 A bifurcation theorem

We shall describe here a general bifurcation theorem which we will use to answer affirmatively the questions posed in the introduction. Our technique will be to reduce the problem, via a new Lyapunov-Schmidt reduction, to one in finite dimensions, where we can easily apply Conley-index techniques. To use these methods directly in the infinite-dimensional case seems too difficult. Indeed, even though the theory has a useful extension to the infinite-dimensional case (Benci 1986; Conley 1978; Rybakowski 1987; Smoller 1983), it is no easy matter to compute changes in indices in this situation; see §4.

In order not to complicate the description, we shall first describe the Lyapunov-Schmidt reduction in the nonequivariant case; the modifications necessary to extend our techniques to the equivariant case will be discussed in the next section.

Let $B_1 \subset B_0 \subset H$, where B_0, and B_1 are Banach spaces, H is a Hilbert space, and the embeddings are assumed to be continuous. (In the applications, we often take $B_1 = C_0^2(D^n)$, $B_0 = C(D^n)$ and $H = L^2(D^n)$, where D^n is an n-ball centred at 0.) Let M be a smooth mapping

$$M : B_1 \times I \to B_0, \quad I = [\lambda_1, \lambda_2] \subset \mathbf{R},$$

and assume that there is a family $\{u_\lambda : \lambda \in I\} \subset B_1$, depending smoothly on λ, which satisfies (1.1). For ease in notation, let

$$dM_\lambda = d_u M(u_\lambda, \lambda).$$

Definition The *peigenspace* P_λ of dM_λ is the closed sub-vector space generated by the set

$$\{v \in B_1 : dM_\lambda v = \mu v, \text{ for some } \mu \geq 0\}.$$

P_λ is thus seen to be the space generated by those eigenvectors of dM_λ corresponding to nonnegative eigenvalues.

If we replace u by $u + u_\lambda$, then with a slight abuse of notation, we write (1.1) as

(2.1) $$M(0, \lambda) = 0, \quad \lambda \in I.$$

For any $\varepsilon > 0$, let $P_{\lambda,\varepsilon}$ be the peigenspace of $dM_\lambda + \varepsilon I$. We assume that there is a fixed $\varepsilon > 0$ such that

(2.2) $$\dim P_{\lambda,\varepsilon} < \infty \quad \forall \lambda \in I.$$

We now state the main result in this section.

Theorem 2.1 *Let M be a gradient operator for each $\lambda \in I$, and assume that M satisfies (2.1) and (2.2). If*

(2.3) $$dM_{\lambda_1} \text{ and } dM_{\lambda_2} \text{ are nonsingular},$$

and

(2.4) $$P_{\lambda_1} \text{ is not isomorphic to } P_{\lambda_2},$$

then there is a λ_0, $\lambda_1 < \lambda_0 < \lambda_2$, for which $(0, \lambda_0)$ is a bifurcation point for $M(0, \lambda) = 0$.

Notice that since $P_\lambda \subset P_{\lambda,\varepsilon}$, condition (2.2) implies that $\dim P_{\lambda_i} < \infty$, $i = 1, 2$; thus (2.4) means that $\dim P_{\lambda_1} \neq \dim P_{\lambda_2}$. We have stated (2.4) in this slightly awkward way in anticipation of the results in the next section where this theorem will be extended to the equivariant case.

Proof. Since M is a gradient for each $\lambda \in I$, it follows that dM_λ is symmetric and so has real eigenvalues. Then (2.2) implies that the eigenvectors of dM_λ corresponding to eigenvalues greater than $-\varepsilon$ lie in a finite-dimensional space. Thus there is a finite-codimensional space $E_\lambda \subset B_1$ such that

$$\langle dM_\lambda e, e \rangle \leq -\varepsilon |e|^2 \quad \forall e \in E_\lambda.$$

(Here the norm and inner product are those in H.) By continuity, there is an open interval J_λ about λ such that

$$\langle dM_\mu e, e \rangle \leq -\frac{\varepsilon}{2} |e|^2 \quad \forall \mu \in J_\lambda, \forall e \in E_\lambda.$$

A finite number of the intervals $J_{\bar{\lambda}_1}, \cdots, J_{\bar{\lambda}_s}$ cover I. Thus let

$$E = \left(\cap_{i=1}^s E_{\bar{\lambda}_i}\right) \cap P_{\lambda_1}^\perp \cap P_{\lambda_2}^\perp;$$

then $E \subset B_1$ has finite codimension. Note that $E_\lambda = P_{\lambda,\varepsilon}^\perp \cap B_1$; thus

$$E = \cap_{i=1}^s P_{\bar{\lambda}_i,\varepsilon}^\perp \cap P_{\lambda_1}^\perp \cap P_{\lambda_2}^\perp \cap B_1 = \left(\cup_{i=1}^s P_{\bar{\lambda}_i,\varepsilon} \cup P_{\lambda_1} \cup P_{\lambda_2}\right)^\perp \cap B_1 \equiv F^\perp \cap B_1,$$

so that E is a closed subspace of B_1 of finite codimension and $P_{\lambda_1} \cup P_{\lambda_2} \subset F$. Now if $\lambda \in I$, then $\lambda \in J_{\bar{\lambda}_k}$ for some k, $1 \leq k \leq s$, and if $e \in F^\perp \cap B_1 = E$, $e \in E_{\bar{\lambda}_k}$ so $\langle dM_\lambda e, e \rangle \leq -\frac{\varepsilon}{2} |e|^2$. Thus dM_λ restricted to $F^\perp \cap B_1$ is uniformly negative definite.

Next, write
$$H = F^\perp \oplus F, \text{ and since } F \subset B_1,$$
$$B_0 = (B_0 \cap F^\perp) \oplus F$$
$$B_1 = (B_1 \cap F^\perp) \oplus F.$$

We can now describe our Lyapunov-Schmidt reduction. Thus for $h \in B_1$, write $h = (x,y)$, $x \in F^\perp$, $y \in F$. Then
$$M(h,\lambda) = M(x,y,\lambda) = (u(x,y,\lambda), v(x,y,\lambda)),$$
where $u \in F^\perp$, and $v \in F$. Since $M(0,0,\lambda) = 0$, we have $u(0,0,\lambda) = 0$. We claim that $u_x \equiv u_x(0,0,\lambda)$, defined on $F^\perp \cap B_1$, is an isomorphism. First, if $u_x \xi = 0$, for some $\xi \neq 0$, then as u_x is strictly negative definite, the inequality $0 > \langle u_x \xi, \xi \rangle$ yields a contradiction. Next, an easy calculation shows that u_x is symmetric. Finally, if $\xi \in H$, we have
$$\langle u_x(\pi_2 \xi), \pi_2 \xi \rangle \leq -\frac{\varepsilon}{2} | \pi_2 \xi |^2,$$
where π_2 = projection onto F. Thus the spectrum of u_x is uniformly bounded away from zero. It follows from these three observations that $u_x(0,0,\lambda)$ is indeed an isomorphism. Therefore the implicit function theorem implies that the equation $u(x,y,\lambda) = 0$ can be solved for $x = x(y,\lambda)$ in a neighbourhood of $(0,0,\lambda)$ of the form $U_\lambda \times \mathcal{O}_\lambda$, where U_λ is a neighbourhood of $(0,0)$ and $\mathcal{O}_\lambda$ is a neighbourhood of λ in I. By compactness, a finite number of the $\mathcal{O}_\lambda$ cover I, say $\mathcal{O}_{\lambda_1}, \cdots, \mathcal{O}_{\lambda_s}$. Therefore we have a unique solution $x = x(y,\lambda)$ on a neighbourhood $U \times I \equiv \cap_1^s U_{\lambda_i} \times I$ of $(0,0,\lambda)$. Now define
$$\phi(y,\lambda) = v(x(y,\lambda), y, \lambda);$$
this is our global Lyapunov-Schmidt reduction.

We next have two lemmas:

Lemma 2.2 *Fix $\lambda \in I$ and consider the ordinary differential equation*

(2.5) $$y_s = \phi(y(s), \lambda)$$

Then this equation is gradient-like.

Proof. If $\psi(x,y)$ is a Lyapunov function for M, then
$$\tilde\psi(y,\lambda) = \psi(x(y,\lambda), \lambda)$$
is easily seen to be a Lyapunov function for (2.5).

Lemma 2.3 $P(d\phi_{(0,\lambda_i)})$ *is isomorphic to* P_{λ_i}, $i = 1,2$.

We omit the somewhat technical proof of this result; see (Smoller and Wasserman 1990b) for the details.

We can now finish the proof of the theorem. To this end, one notes first that solutions of $M = 0$ correspond in a one-one way to solutions of $\phi = 0$. One then checks that (2.2) implies that $d\phi_{(0,\lambda_i)}$ is nonsingular for $i = 1, 2$. It follows that the Conley indices for the rest points $(0, \lambda_i)$ of the equation (2.5) satisfy $h(0, \lambda_1) \neq h(0, \lambda_2)$. Thus if $\mathcal{V}$ is any neighbourhood of 0, (in the finite-dimensional space), then 0 cannot be the maximal invariant set in $\mathcal{V}$ for each $\lambda \in I$. Thus there is a point $\lambda(\mathcal{V}) \in I$ for which 0 is not the maximal invariant set in $\mathcal{V}$ for the equation $y_s = \phi(y(s), \lambda(\mathcal{V}))$. The maximal invariant set in $\mathcal{V}$ for this equation must then contain a point $y \neq 0$. The gradient-like nature of the equation forces the α- and ω-limit sets of y to differ. Hence the equation admits another rest point in $\mathcal{V}$, different from 0. Now let the neighbourhoods $\mathcal{V}$ shrink to 0. The $\lambda(\mathcal{V})$'s have a convergent subsequence $\lambda_i \to \lambda_0$, and $(0, \lambda_0)$ is easily seen to be a bifurcation point for $\phi(y, \lambda) = 0$, (so $\lambda_1 < \lambda_0 < \lambda_2$), and hence for $M = 0$.

3 Equivariant Conley index

We now show how to extend the Conley index ideas to the case where there is a group acting on the space. This will give us a finer index invariant which will prove quite useful in applications to symmetry-breaking problems. Again let B_0 and B_1 be Banach spaces, let H be a Hilbert space, and assume $B_1 \subset B_0 \subset H$, where the embeddings are continuous. Let $I = [\lambda_1, \lambda_2]$ be an interval in $\mathbf{R}$, and let M be a smooth mapping $M : B_1 \times I \to B_0$. We again consider solutions of the equation $M(u, \lambda) = 0$.

We assume the existence of a compact Lie group G acting on H with B_0 and B_1 invariant (so $gu \in B_1$ and $gv \in B_0$ for all $g \in G$, $u \in B_1$, $v \in B_0$). We further assume that M is *equivariant* with respect to G, in the sense that for all $u \in B_1$, $\lambda \in I$, $g \in G$, we have

$$M(gu, \lambda) = gM(u, \lambda).$$

Let (u_λ, λ) be a smooth curve of *invariant* (symmetric) solutions of $M = 0$; thus

(3.1) $$M(u_\lambda, \lambda) = 0, \forall \lambda \in I,$$

and $gu_\lambda = u_\lambda$ for all $g \in G$ and all $\lambda \in I$. Now as we have discussed in the introduction, each peigenspace P_λ is a representation of G. Since G is compact, it follows by the Peter-Weyl Theorem, (see, e.g., (Knapp 1986)), that each P_λ is a sum of finite-dimensional irreducible[4] representations.

Again write $dM_\lambda \equiv d_u M(u_\lambda, \lambda)$, and as before, we assume that there is an $\varepsilon > 0$ such that the peigenspaces $P_{\lambda,\varepsilon}$ of $dM_\lambda + \varepsilon I$ satisfy (2.2).

Theorem 3.1 *Let M be a gradient operator for each $\lambda \in I$, and assume that M satisfies (2.2) and (3.1). If*

(3.2) $$dM_{\lambda_1} \text{and } dM_{\lambda_2} \text{ are nonsingular,}$$

---

[4] A representation V of G is called irreducible if V has no proper, closed, invariant (under all operators T_g, $g \in G$) subspaces.

and

(3.3) P_{λ_i} contains k_i copies of a fixed irreducible representation of G, where $k_1 \neq k_2$,

then there is a λ, $\lambda_1 < \lambda < \lambda_2$ such that (u_λ, λ) is a bifurcation point.

The theorem is proved by extending the Conley index to the equivariant case, obtaining a G-Conley index $h_G(\cdot)$. This puts more structure on the Conley index and thus makes it more useful. For example, it is possible to have $h(I_1) = h(I_2)$ but $h_G(I_1) \neq h_G(I_2)$. Thus we can prove bifurcation theorems using the G-Conley index h_G, while no such statement is possible using the (usual) Conley index, h. Also, in the finite-dimensional case, (as in the nonequivariant case), at a nondegenerate critical point I, the Conley index of I reduces to the Morse index of I, i.e., we have an equivariant homotopy equivalence

$$h(I) \approx D(V)/S(V).$$

Here V is the peigenspace of I and $D(V)$ (resp. S(V)), denotes the unit ball (resp. unit sphere) in V. Now at distinct parameter values $\lambda_1 \neq \lambda_2$, we can have dim $V_1 =$ dim V_2 and thus $h(I_1, \lambda_1) = h(I_2, \lambda_2)$. But by a result of Lee and Wasserman(1975), if $G = O(n)$ (for example), and $V_1 \neq V_2$ as representations of $O(n)$, then $h_{O(n)}(I_1, \lambda_1) \neq h_{O(n)}(I_2, \lambda_2)$. It is thus of some interest to extend the Conley index to the equivariant case. We shall now outline such an extension.

Let X be a metric space, and G a compact Lie group. Suppose that ψ is a G-flow[5] on X, i.e.,

$$\psi : (\mathbf{R} \times G) \times X \to X,$$

ψ is continuous, and satisfies $\forall x \in X$, $\forall g_1, g_2 \in G$, $\forall t_1, t_2 \in \mathbf{R}$,

$$\psi(0, e, x) = x \quad (e = id. \text{ in } G),$$

and

$$\psi(t_1, g_1, \psi(t_2, g_2, x)) = \psi(t_1 + t_2, g_1 g_2, x).$$

If $g = e$, $\psi_t(x) \equiv \psi(t, e, x)$ defines a flow on X, and if $t = 0$, ψ induces an action of G on X by

$$gx \equiv \psi(0, g, x).$$

Let X/G denote the *orbit space* of X with respect to G, i.e., X/G is the set of orbits in X under G. Thus, we define an equivalence relation on X by $x \sim y$ if $y = gx$ for some $g \in G$; then X/G is the set of equivalence classes. Let $\pi : X \to X/G$ be the canonical projection. We put the quotient topology on X/G, i.e., $A \subset X/G$ is open if $\pi^{-1}(A)$ is open in X. Then if $(\bar{N}_1, \bar{N}_2)$ is an index pair in X/G, $(\pi^{-1}(\bar{N}_1), \pi^{-1}(\bar{N}_2))$ is an index pair in X(see Pacella 1986). Moreover, if (N_1, N_2) is a G-invariant index pair in X, then $(\pi(N_1), \pi(N_2))$ is an index pair in X/G. If I is a G-invariant set in X, and I is also an isolated invariant set for the flow ψ_t, then $\pi(I)$ is an isolated

---

[5] Or semiflow, (or local flow), in which case $\mathbf{R}$ is to be replaced by $\mathbf{R}_+$.

invariant set for the induced flow on X/G, and we define $h_G(I)$, the *G-Conley Index* of I, to be the equivariant homotopy type of the pointed space

$$h_G(I) = \pi^{-1}(\bar{N}_1)/\pi^{-1}(\bar{N}_2),$$

where $(\bar{N}_1, \bar{N}_2)$ is an index pair for $\pi(I)$. [Two spaces X and Y are said to be of the same equivariant homotopy type with respect to G if they are homotopy equivalent, and the group action commutes with all of the relevant maps. In other words there exist maps $f: X \to Y$, and $h: Y \to X$ such that $f \circ h$ and $h \circ f$ are homotopic to the identity through equivariant maps, i.e., maps which commute with the group action. It is also not too hard to show that there is an induced flow on X/G given by $[x] \cdot t = [x \cdot t]$ and if I/G is an isolated invariant set for the flow on X/G, and $(\bar{N}_1, \bar{N}_0)$ is an index pair for I/G, then $(\pi^{-1}(\bar{N}_1), \pi^{-1}(\bar{N}_0))$ is a G-invariant index pair for I.]

Let (N_1, N_0) and $(\tilde{N}_1, \tilde{N}_0)$ be G-invariant index pairs for $I \subset X$. Then there exist equivariant maps f and h (i.e., maps which commute with the G-action), satisfying

$$f: (N_1/N_0, N_0) \to (\tilde{N}_1/\tilde{N}_0, \tilde{N}_0)$$
$$h: (\tilde{N}_1/\tilde{N}_0, \tilde{N}_0) \to (N_1/N_0, N_0),$$

such that both $h \circ f$ and $f \circ h$ are equivariantly homotopic to the identity map; i.e., the pointed spaces $(N_1/N_0, N_0)$ and $(\tilde{N}_1/\tilde{N}_0, \tilde{N}_0)$ are of the same equivariant homotopy type. We can thus unambiguously define the G-invariant Conley index of I to be this equivariant homotopy type. Note that this definition reduces to the usual one when G is the trivial group. The definition given here is richer since as the relevant spaces admit a group action, we can distinguish indices which have the same homotopy type as pointed spaces, but are not of the same homotopy type as pointed G-spaces.

Definition If dM_λ is nonsingular for some $\lambda \in I$, the *reduced G-index* of u_λ, $h_G^R(u_\lambda, \lambda)$, is the equivariant homotopy type of the pointed G-space:

$$h_G^R(u_\lambda, \lambda) = D(P_\lambda)/S(P_\lambda).$$

Note that in the finite-dimensional case, this definition agrees with our equivariant formulation of the Conley index as given above. We now have

Theorem 3.2 *With M as defined above, (satisfying (2.2) and (3.1)), assume that M is a gradient for each $\lambda \in I$. Suppose that*

(3.4) $\qquad\qquad\qquad dM_{\lambda_i}$ is nonsingular for $i = 1, 2,$

and [6]

(3.5) $\qquad\qquad\qquad h_G^R(u_{\lambda_1}, \lambda_1) \not\approx h_G^R(u_{\lambda_2}, \lambda_2).$

Then there exists λ, $\lambda_1 < \lambda < \lambda_2$ such that (u_λ, λ) is a bifurcation point.

---

[6] $A \approx B$ means that A and B are of the same equivariant homotopy type.

The theorem is proved in exactly the same way as in the nonequivariant case (Theorem 2.1). The connection of hypothesis (3.5) with hypothesis (3.3) in Theorem 3.1 follows from results in (Lee and Wasserman 1975). Indeed, we have the following theorem which is applicable to a wide class of groups G; we only give the statement for $G = O(n)$.

Theorem 3.3 *If $G = O(n)$ and V and W are representations of $O(n)$, then the pointed $O(n)$-spaces $D(V)/S(V)$, $D(W)/S(W)$ are equivariantly homotopy equivalent if and only if V is isomorphic to W as $O(n)$ representations.*

Since $h_{O(n)}^R(u_{\lambda_i}, \lambda_i) = D(P_{\lambda_i})/S(P_{\lambda_i})$, we see that Theorems 3.1 and 3.2 are equivalent. The importance of this statement lies in the fact that one can often verify (3.3) via a study of the linearized equations; this will be demonstrated for an interesting example in the next section.

4 Application to semilinear elliptic equations

In this section we shall apply our general bifurcation theorems to study the symmetry-breaking problem for solutions of the semilinear elliptic equation

(4.1) $$\Delta u(x) + f(u(x)) = 0, \ x \in D_R^n,$$

with boundary conditions

(4.2) $$\alpha u(x) - \beta du(x)/dn = 0, x \in \partial D_R^n.$$

Here D_R^n denotes an n-ball of radius R, f is a smooth function, $\alpha^2 + \beta^2 = 1$ and d/dn denotes differentiation in the radial direction.

We make the following assumptions on f:
There exist points $b < 0 < \gamma$ such that

$$(H) \begin{cases} (H_1) & f(\gamma) = 0, f'(\gamma) < 0 \\ (H_2) & F(\gamma) > F(u) \text{ if } b < u < \gamma \text{ (here } F' = f \text{ and } F(0) = 0) \\ (H_3) & F(b) = F(\gamma) \\ (H_4) & \text{if } f(b) = 0, \text{ then } f'(b) < 0 \\ (H_5) & uf(u) + 2(F(\gamma) - F(u)) > 0 \text{ if } b < u < \gamma. \end{cases}$$

We refer the reader to (Smoller and Wasserman 1990a) for a discussion of these conditions; we merely point out that (H_2) and (H_5) both hold if $uf(u) > 0$ for $u \in (b,\gamma)/\{0\}$; for example if $f(u) = u(1-u)$, then (H) holds.

For equation (4.1), the relevant symmetry group is $G = O(n)$, i.e., $O(n)$ is the largest group which leaves (4.1) invariant. Solutions of (4.1) which possess this maximal symmetry are the radial solutions—any nonradial, (i.e., asymmetric) solution is called a *symmetry-breaking solution*. Now radial solutions $u = u(r)$ of (4.1) satisfy the equation

(4.3) $$u'' + \frac{n-1}{r}u' + f(u) = 0, \ 0 < r < R,$$

(where $r = |x|$, and $u = u(r)$), together with the boundary conditions

(4.4) $$u'(0) = 0 = \alpha u(R) - \beta u'(R).$$

The solution of (4.3) which satisfies the initial conditions $u(0) = p > 0$, $u'(0) = 0$, will be denoted by $u(r,p)$, and p will be considered as a parameter. We write

$$\theta_0 = \tan^{-1}(\alpha/\beta), -\pi/2 \leq \theta_0 < \pi/2, \text{ and}$$

$$\theta(r,p) = \tan^{-1}(u'(r,p)/u(r,p)).$$

For k a given nonnegative integer, and $f(p) > 0$, we define the function $p \to T_k(p)$, (the "time map") by

(4.5) $$\theta(T_k(p), p) = \theta_0 - k\pi.$$

We have shown in (Smoller and Wasserman 1989), that T_k is well-defined for p near γ. Observe that $T_k(p)$ now plays the role of R, and thus R varies with p. A solution of (4.3) ,(4.4) will be said to belong to the kth *nodal class* if (4.5) holds. Since k will be fixed throughout this section, we shall write $T_k(p)$ as $T(p)$.

Next, since $O(n)$ is a compact Lie group, our previous remarks apply so that every irreducible representation of $O(n)$ is finite-dimensional, and every finite-dimensional representation is a direct sum of its irreducible representations (cf footnote 4). Moreover, the eigenspaces E_N of the (covariant) Laplacian on S^{n-1} are irreducible representations of $O(n)$; (a proof of this fact is given in the appendix). These eigenspaces E_N, corresponding to the eigenvalue $\lambda_N = -N(N+n-2)$, have dimension

$$\ell_N = \binom{N+n-2}{N} \frac{2N+n-2}{N+n-2} = O(N^{n-2}).$$

In order to illustrate how representation theory is used in our symmetry-breaking problem, we must discuss some results in (Smoller and Wasserman 1990a). Thus it was shown that there is a $\sigma > 0$ such that the interval $(\gamma - \sigma, \gamma)$ lies in the domain of T. For $p \in (\gamma - \sigma, \gamma)$, consider the linearized operator

$$L^p(w) = \Delta w + f'(u(\cdot,p))w, \ |x| < T(p),$$

where

$$w \in \Phi_p = \{\phi \in C^2(|x| < T(p)) : \alpha\phi(x) - \beta d\phi(x)/dn = 0, |x| = T(p)\}.$$

Since any function on the ball has a spherical harmonic decomposition, we may write
$$w(r,\theta) = \sum_{N\geq 0} a_N(r)\Phi_N(\theta),\ 0 \leq r \leq T(p),\ \theta \in S^{n-1},$$
where S^{n-1} denotes the unit sphere in $\mathbf{R}^n$, and $\Phi_N \in E_N$. This leads to the following decomposition of L^p as
$$L^p = \oplus \sum_{N\geq 0} L^p_N,$$
where each L^p_N is defined for $\phi \in \Phi_p$ by

(4.6) $\quad L^p_N \phi = \phi'' + \dfrac{n-1}{r}\phi' + \left(f'(u(\cdot,p)) + \dfrac{\lambda_N}{r^2}\right)\phi,\ 0 < r < T(p).$

Now it was proved in (Smoller and Wasserman 1990a) that there exists a sequence $q_N \uparrow \gamma$ such that for large N, say $N \geq N_0$, $L^{q_N}_N$ has positive spectra; i.e., there exist positive numbers $\mu_1^N, \mu_2^N, \cdots, \mu_{k_N}^N$, and nontrivial functions $a_j^N \in \Phi_{q_N}$, $j = 1, 2, \cdots, k_N$, such that
$$L^{q_N}_N a_j^N = \mu_j^N a_j^N, j = 1, 2, \cdots, k_N.$$

Under the additional hypothesis that f is analytic[7] we may assume that each radial solution $u(\cdot, q_N)$ is nondegenerate in the sense that $0 \notin sp(L^{q_N})$. Moreover, we may also assume that $L^{q_N}_M$ has positive spectra if $M > N$. Using the fact that the λ'_Ns decrease monotonically, it is not difficult to show that whenever L^p_N has a positive eigenvalue, the same is true for each operator L^p_M, $M < N$, see (Smoller and Wasserman 1990a). It follows that the peigenspace of each $u(\cdot, q_N)$ is a representation of $O(n)$ of the form
$$k_N E_N \oplus \cdots \oplus k_1 E_1 \oplus k_0 E_0,$$
where each k_j is a nonnegative integer, $j = 0, \cdots, N$. Therefore the dimension of the peigenspace of $u(\cdot, q_N)$ is
$$\dim P(u(\cdot, q_N)) = k_N \ell_N + \cdots + k_1 \ell_1 + k_0.$$

Thus it is a priori possible that $\dim P(u(\cdot, q_N)) = \dim P(u(\cdot, q_{N+1}))$; i.e.,
$$k_N \ell_N + \cdots + k_1 \ell_1 + k_0 = k'_{N+1}\ell_{N+1} + k'_N \ell_N + \cdots + k'_1 \ell_1 + k'_0,$$
so that the corresponding peigenspaces, P_N, P_{N+1} of $u(\cdot, q_N)$ and $u(\cdot, q_{N+1})$ are isomorphic. That is, the homotopy type of the corresponding pointed spaces satisfy
$$D(p_N)/S(p_N) \approx D(p_{N+1})/S(p_{N+1}).$$

Thus if we do not take account of the group structure we cannot apply Theorem 2.1. Thus we cannot use the ordinary(i.e., nonequivariant) Conley index to prove that

---

[7] This assumption is needed only if $k > 1$, where k denotes the nodal class of radial solutions.

bifurcation occurs - there is no index change at $u(\cdot, q_N)$ and $u(\cdot, q_{N+1})$. On the other hand, P_N and P_{N+1} differ as representations of $O(n)$, because P_N contains no copies of E_{N+1}. Thus Theorem 3.1 is applicable and shows that for each $N \geq N_0$, there is a point p_N, $q_N < p_N < q_{N+1}$ for which $u(\cdot, p_N)$ is a bifurcation point. Now as we have shown in (Smoller and Wasserman 1989), under the given hypotheses on f, for p near γ, no radial bifurcation is possible on $u(\cdot, p)$ if $T'(p) \neq 0$. Since it was proved in (Smoller and Wasserman 1989) that $T'(p) > 0$ if p is near γ, it follows that for all sufficiently large N the symmetry breaks on $u(\cdot, p_N)$. We thus have the following theorem.

Theorem 4.1 *Let f satisfy hypotheses (H) and let k be a given integer representing a fixed nodal class of radial solutions of (4.1), (4.2). Assume that f is analytic if $k > 1$. Then there exists a sequence of points $p_N \uparrow \gamma$ such that the symmetry breaks on each radial solution $u(\cdot, p_N)$.*

5 Concluding remarks

The question of the structure of the set of bifurcating asymmetric solutions is not fully resolved. We merely mention that we have previously shown that at each symmetry-breaking bifurcation point there bifurcates out a family of $O(n-1)$-invariant solutions; i.e., axi-symmetric solutions. Moreover, we can prove that there are degenerate radial solutions for which there also bifurcates out (among others), distinct solutions having symmetry groups (at least) $O(p) \times \tilde{O}(n-p)$, where $O(p)$ and $\tilde{O}(n-p)$ are $n \times n$ orthogonal matrices of the forms

$$O(p) = \begin{pmatrix} * & 0 \\ 0 & I_{n-p} \end{pmatrix}, \quad \tilde{O}(n-p) = \begin{pmatrix} I_p & 0 \\ 0 & * \end{pmatrix},$$

respectively. For a proof of this statement see (Smoller and Wasserman 1990b).

Next, we want to point out an interesting technical difficulty which arises in studying the symmetry-breaking problem for solutions of (4.1), (4.2). Namely, since we have taken p as a parameter, the radius R of the ball on which the radial solution $u(\cdot, p)$ is defined, depends on p; i.e., $R = T(p)$. The space

$$\{(\phi, p) \in C^2(0, T(p)) \times \mathbf{R}_+ : \alpha \phi(x) - \beta \frac{d\phi(x)}{dn} = 0, \; |x| = T(p)\}$$

on which our bifurcation is presumed to occur is not in the form of a product space $B \times \Lambda$, where B is a Banach space and Λ is the parameter space. If we change variables, writing $y = x/R$, then (4.1), (4.2) becomes

$$\Delta u(y) + R^2 f(u(y)) = 0, \; |y| \leq 1$$

$$\alpha u(y) - \beta R du(y)/dn = 0, \; |y| = 1.$$

Thus, if $\alpha\beta = 0$ (i.e., Dirichlet or Neumann boundary conditions), then the parameter R doesn't appear in the boundary conditions, and we do have the desired product structure. However, if $\alpha\beta \neq 0$, then we again do not have the relevant product structure. This problem is overcome by showing that the space

$$\{(u,\lambda) \in C^2(0,1) \times \mathbf{R}_+ : \alpha u(y) - \beta\lambda\frac{du(y)}{dn} = 0, \mid y \mid = 1\}$$

forms a vector bundle over $\mathbf{R}_+$, and is thus locally a product space. This local product structure is sufficient for doing bifurcation theory; for details see (Smoller and Wasserman 1990b).

Finally, we mention that the Lyapunov-Schmidt reduction to the peigenspace, as described in §2 is a framework on which the application of the Conley Index Theory to reaction-diffusion equations should be performed. In particular, the development in (Smoller 1983, ch24 §D) should be done in this setting. This subtle point was carelessly glossed over in (Smoller 1983).

Appendix

We prove here the following apparently well-known result whose proof we have been unable to locate in the literature.

Theorem 5.1 *Let V be an eigenspace of the Laplace operator on the unit sphere $S \subset \mathbf{R}^n$. Then V is an irreducible representation of $O(n)$.*

Proof. Since the action of $O(n)$ commutes with the Laplace operator on S, V is easily seen to be a representation of $O(n)$. We must show that V is irreducible.

Equip V with the L_2-inner product, with respect to the $O(n)$–invariant measure on S. We will need the following two facts about V:

(i) If $f \in V$, and f vanishes in a neighbourhood of some point, then $f|_S \equiv 0$.

(ii) If $f \in V$ and f is invariant under all those transformations $T \in O(n)$ which fix a specific $\xi \in S$, and $f(\xi) = 0$, then $f|_S \equiv 0$.

(If we recall that any $f \in V$ is the restriction to S of a homogeneous, harmonic polynomial on $\mathbf{R}^n$, then (i) is obvious. To show (ii), we use the following result (Müller 1966): Let $\xi \in S$; then there is a unique $\phi \in V$ which satisfies $\phi(\xi) = 1$, and is invariant under $\{T \in O(n) : T\xi = \xi\}$.[8] Assuming this, let $f \in V$ satisfy the hypotheses of (ii). Let ϕ be as in the quoted theorem, for this given ξ. Since the quoted result implies that $\phi - f \equiv \phi$, we have $f \equiv 0$).

Having established (i) and (ii), we prove the theorem. Thus suppose that the $O(n)$-representation V were reducible; let X be a nontrivial $O(n)$-invariant subspace. Since $O(n)$ preserves the L_2-inner product, $X^\perp$ is also a nontrivial invariant subspace. Pick nonzero elements $f \in X$, $g \in X^\perp$. Choose $\eta \in S$ such that $f(\eta) \neq 0$;

---

[8] In (Müller 1966), ξ is a given basis vector ε, but one can remove this restriction by choosing $A \in O(n)$ such that $A\xi = \varepsilon$, and proceeding in a straightforward manner.

then f is nonzero in a neighbourhood of η and by (i), there is a ξ in this neighbourhood with $g(\xi) \neq 0$. Thus $f(\xi)g(\xi) \neq 0$. Multiplying by constants, we may assume $f(\xi) = 1 = g(\xi)$. Now symmetrize f and g about ξ, i.e., define

$$\bar{f}(\alpha) = \int_{\substack{T \in O(n) \\ T(\xi) = \xi}} f(T\alpha)dT,$$

the integral being taken over the subgroup of $O(n)$ which fixes ξ, and is with respect to normalized Haar measure on this subgroup. Thus $\bar{f}$ is invariant under this subgroup and $\bar{f}(\xi) = 1$. Similarly, we obtain $\bar{g}$. Now since $f \in X$ and X is $O(n)$-invariant, $\bar{f} \in X$; similarly $\bar{g} \in X^\perp$; thus $\langle \bar{f}, \bar{g} \rangle = 0$. By (ii), $\bar{f} - \bar{g} = 0$, so that $0 = \langle \bar{f}, \bar{f} \rangle = |\bar{f}|^2$. Thus $\bar{f} \equiv 0$, and this contradicts $\bar{f}(\xi) = 1$.

References

[1] Amann, H., and Zehnder, E.(1980). Nontrivial solutions for a class of nonresonance problems and applications to nonlinear differential equations. *Annali della Scuola Normale Superiore di Pisa*, **7**, 539-603 .

[2] Benci, V.(1986). A generalization of the Conley-index theory. *Rendiconti di Matematica, Universita di Trieste*, **18**, 6-39.

[3] Conley, C.(1978). *Isolated invariant sets and the Morse index*. C.B.M.S., No. 38, A.M.S., Providence.

[4] Chow, B.N., and Lauterbach, R.(1985) *A bifurcation theorem for critical points of variational problems*. IMA report, Minneapolis.

[5] Kielhofer, H.(1988). A bifurcation theorem for potential operators. *Journal of Functional Analysis*, **77**, 1-8.

[6] Knapp, A.W.(1986). *Representation theory of semisimple groups*. Princeton University Press, Princeton.

[7] Lee, C.N., and Wasserman, A.(1975). On the groups $JO(G)$. Memoirs of American Mathematical Society, **159**, A.M.S., Providence.

[8] Müller, C.(1966). *Spherical harmonics*. Springer Lecture Notes in Mathematics. No. 17, Springer-Verlag, Berlin.

[9] Pacella, F.(1986). Equivariant Morse theory for flows and an application to the N-body problem. *Transactions of the American Mathematical Society*, **297**, 41-52.

[10] Rybakowski, K.(1987). *The homotopy index and partial differential equations*. Springer-Verlag, Berlin.

[11] Smoller, J.(1983). *Shock waves and reaction-diffusion equations.* Springer-Verlag, New York.

[12] Smoller, J., and A. Wasserman.(1989). On the monotonicity of the time map. *Journal of Differential Equations,* **77**, 287-303.

[13] Smoller, J., and Wasserman, A.(1990a). Symmetry, degeneracy, and universality in semilinear elliptic equations. To appear in *Journal of Functional Analysis.*

[14] Smoller, J., and Wasserman, A.(1990b). Bifurcation and symmetry-breaking. To appear in *Inventiones Mathematicae.*

Remarks on the equilibrium theory for the Cahn-Hilliard equation in one space dimension

Nicholas D. Alikakos [*]
University of Tennessee

William R. McKinney
North Carolina State University

0 Introduction

The Cahn-Hilliard equation in one space dimension equipped with no flux boundary conditions is given by

(CH)
$$u_t = (-\varepsilon^2 u_{xx} + W'(u))_{xx}, \quad -1 < x < 1$$
$$u_x = u_{xxx} = 0 \text{ at } x = -1, 1.$$

Equation (CH) is one of the most celebrated models devised to study the phenomena of separation and coarsening for a melted binary alloy under conditions of constant temperature, where the unknown function u is the concentration of one of the components (Cahn 1961; Cahn and Hilliard 1958). In this article we are concerned with the equilibrium theory for (CH) (Van der Waals 1893), i.e., the study of the stable time independent solutions of (CH). In particular we study the behaviour of these solutions for ε small.

1 Statement of results

Set for $\varepsilon \geq 0$

$$J_\varepsilon(u) = \int_{-1}^{1} \left[\frac{\varepsilon^2}{2}(u_x)^2 + W(u) \right] dx$$

where W is a smooth double-well function as in Figure 1, $W \geq W(\alpha) = W(\beta) = 0$, and consider the variational problem

(1.1) $$\min J_\varepsilon \text{ over } W^{1,2}(-1, 1)$$

subject to the integral (mass) constraint

$$\int_{-1}^{1} u \, dx = M$$

---

[*]Partially supported by NSF DMS-8804631 and the Science Alliance, a program of the Tennessee Centers of Excellence.

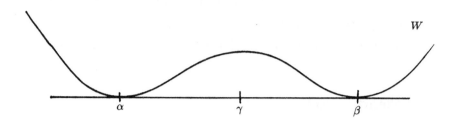

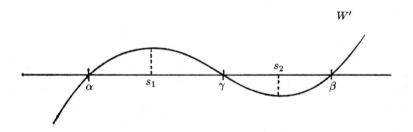

Figure 1. Graphs of the functions W and W'

where M is a specified constant satisfying the condition

$$\alpha < \frac{M}{2} < \beta.$$

In their fundamental paper Carr, Gurtin and Slemrod (1984) established the following result:

Theorem 1.1 *(i) For $\varepsilon > 0$ small enough, (1.1) has a global minimizer u_ε, unique up to a reversal ($x \to -x$);*
(ii) u_ε is strictly monotone;
(iii) as $\varepsilon \to 0$, u_ε (or its reversal) approaches the monotonically decreasing single-interface solution of the variational problem

(1.2) $$\min J_0$$

over the class of functions u that belong to $L^1(-1,1)$ with the property that $W(u)$ also belongs to $L^1(-1,1)$;

(iv) u_ε satisfies the (Euler-Lagrange) equation

(EL)
$$\begin{aligned} \varepsilon^2 u'' &= W'(u) - \sigma_\varepsilon, \quad -1 < x < 1, \\ u'(-1) &= u'(+1) = 0 \end{aligned}$$

and the constant σ_ε (Lagrange multiplier) satisfies the estimate

(EXP) $\quad \sigma_\varepsilon = O\left(e^{-\frac{C}{\varepsilon}}\right)$

for some positive constant C.

The proof of part (iii) of the above theorem is given in (Carr, Gurtin and Slemrod 1984) via a remarkable phase plane argument in the course of which estimate (EXP) is also established. Part (ii) on the other hand is established by an independent, relatively simple, argument(Carr, Gurtin and Slemrod 1984, Theorem 8.2). Subsequently it was observed by N. Owen (personal communication) and independently by Alikakos and Shaing(1987) that by making use of the uniform estimate

Total Variation of $u_\varepsilon \leq$ Const.,

which follows easily from part (ii), one can establish (iii) in a different way. In this article, continuing in the vein of (Alikakos and Shaing 1987) we give a PDE proof for (EXP). In fact we show that

(1.3) $\quad\quad\quad\quad \sigma_\varepsilon = O\left(e^{-\frac{2\nu}{\varepsilon}d}\right)$

where d is the distance of the layer from the boundary of the interval and ν is a number that can be chosen as close to $\min\{W'''(\alpha), W'''(\beta)\}$ as desired. Also in the course of proving (1.3) we refine statement (iii) of the theorem in various ways(see estimates (4.31), (4.32), (4.54), (4.55) and (4.57), (4.58)). We should mention that our argument does not depend very much on the specific form of the functions J_ε ($W(u)$ could be replaced by $W(u,x)$, the gradient term by a more general penalty term, etc., $\cdots$) and therefore it may be useful in other situations. Finally in our considerations it is not essential to deal with global minimizers. In fact, after reproducing for the convenience of the reader the proof of (iii) from (Alikakos and Shaing 1987), we supply the necessary few modifications needed for establishing the following for families of *local minimizers*.

Proposition *Let $\{u_\varepsilon\}$ be a family of local minimizers of problems (1.1) with the additional proviso that*

(*) $\quad\quad\quad s_1 < \frac{M}{2} < s_2$ *(see Figure 1).*

Then as $\varepsilon \to 0$, u_ε (or its reversal) approaches the monotonically decreasing single-interface solution of the problem (1.2). Moreover all the estimates given in this paper hold for this class of minimizers.

Hypothesis (∗) excludes the possibility of the family $\{u_\epsilon\}$ converging to a stable uniform state outside the spinodal region.

For part (ii), which we will assume in our argument, we refer the reader to (Carr, Gurtin and Slemrod 1984, p348-349) for a self-contained proof.

The analog of (1.3) in higher space dimensions is quite different and has been established for global minimizers in (Luckhaus and Modica, preprint).

Following (Gurtin 1985) we will take W a C^2 function with exactly three critical points, $\alpha < \gamma < \beta$, with α, β local minima and γ local maximum and

(1.4)
$$W \geq 0, \; W(\alpha) = W(\beta) = 0$$
$$W''(\alpha), \; W''(\beta), \; -W''(\gamma) > 0$$

Also M is a given constant with

$$\alpha < \frac{M}{2} < \beta.$$

Acknowledgement

We are indebted to Peter Bates for suggesting a number of changes that improved a former version of this article. We are particularly grateful for his argument in the proof of the proposition that eliminated an extra hypothesis.

2 Proof of (iii)

First without loss of generality we assume that the family of $\{u_\epsilon\}$ consists of monotonically decreasing functions. We split the argument into steps.

1. $|u_\epsilon|_{L^\infty} \leq$ Const.

Set
$$\beta_\epsilon = u_\epsilon(-1), \; \alpha_\epsilon = u_\epsilon(+1).$$

By our hypothesis
$$\max_{-1 \leq x \leq 1} u_\epsilon(x) = \beta_\epsilon, \quad \min_{-1 \leq x \leq 1} u_\epsilon(x) = \alpha_\epsilon,$$

and since $u'_\epsilon(-1) = u'_\epsilon(+1) = 0$, we have

(2.1)
$$u''_\epsilon(-1) \leq 0, \; u''_\epsilon(+1) \geq 0.$$

Therefore from equation (EL)

(2.2)
$$0 \geq W'(\beta_\epsilon) - \sigma_\epsilon, \; W'(\alpha_\epsilon) - \sigma_\epsilon \geq 0$$

and thus

(2.3)
$$W'(\alpha_\epsilon) \geq W'(\beta_\epsilon).$$

Taking into account the shape of W' we obtain the desired estimate.

2. $|\sigma_\varepsilon| \leq \text{Const.}$
This is immediate from (2.2) above and the estimate of step 1.

3. $\int_{-1}^{+1} (u'_\varepsilon)^2 dx \leq \frac{\text{Const.}}{\varepsilon}$
Let

(2.4) $$x_0 = \frac{M - (\alpha + \beta)}{\beta - \alpha}, \quad -1 < x_0 < 1,$$

and set

(2.5) $$\tilde{u}_\varepsilon(x) = \begin{cases} \beta & \text{on } [-1, x_0 - \varepsilon] \\ \alpha & \text{on } [x_0 + \varepsilon, 1] \\ \text{linear interpolant} & \text{on } [x_0 - \varepsilon, x_0 + \varepsilon]. \end{cases}$$

Note that

(2.6) $$\int_{-1}^{+1} \tilde{u}_\varepsilon dx = M, \quad J_\varepsilon(\tilde{u}_\varepsilon) \leq (\text{Const.})\varepsilon$$

and therefore, since u_ε is a global minimizer,

(2.7) $$J_\varepsilon(u_\varepsilon) \leq J_\varepsilon(\tilde{u}_\varepsilon) \leq (\text{Const.})\varepsilon.$$

From this inequality the desired estimate follows, and moreover we obtain that

(2.8) $$\lim_{\varepsilon \to 0} J_\varepsilon(u_\varepsilon) = 0.$$

4. As $\varepsilon \to 0$,
$$u_\varepsilon \to \begin{cases} \beta & \text{on } [-1, x_0) \\ \alpha & \text{on } (x_0, 1] \end{cases}$$

and
$$\sigma_\varepsilon \to 0$$

as $\varepsilon \to 0$.

For establishing this we multiply (EL) by ϕ, a smooth test function with compact support in $[-1, +1]$, and integrate by parts to obtain

(2.9) $$-\varepsilon^2 \int_{-1}^{+1} u'_\varepsilon \phi' = \int_{-1}^{+1} W'(u_\varepsilon) \phi dx - \sigma_\varepsilon \int_{-1}^{+1} \phi dx.$$

By Helly's compactness theorem (Natanson 1961, p220) along a subsequence

(2.10) $$u_\varepsilon \to \bar{u} \quad \text{pointwise on } [-1, 1]$$

as $\varepsilon \to 0$. Taking possibly a further subsequence we may assume that

$$\sigma_\varepsilon \to \bar\sigma$$

as $\varepsilon \to 0$ (by the estimate in step 2). We now take the limit of (2.9) along the subsequence. By step 3 the left hand side converges to zero and so by (2.10) and the L^∞ bound established in step 1 we obtain

$$\int_{-1}^{+1} W'(\bar u)\phi dx = \bar\sigma \int_{-1}^{+1} \phi dx,$$

and since ϕ is arbitrary

(2.11) $$W'(\bar u) = \bar\sigma.$$

Assume for the moment that $\bar\sigma \neq 0$. Then from Figure 1 we see that $\bar u$ can take at most three values $\{r_1, r_2, r_3\}$, where the r_i's are roots to the equation

(2.12) $$W'(r) - \bar\sigma = 0.$$

By (2.8), (2.10) and the L^∞ estimate of step 1 we obtain

(2.13) $$\int_{-1}^{+1} W(\bar u)dx = \lim_{\varepsilon \to 0}\int_{-1}^{+1} W(u_\varepsilon)dx = 0.$$

This statement is in contradiction with the assumption $\bar\sigma \neq 0$ as can be seen via (2.12). Therefore $\bar\sigma = 0$ and $\{r_1, r_2, r_3\} = \{\alpha, \beta, \gamma\}$. Using (2.13) once more we see that $\bar u$ cannot take the value γ on an open interval and thus, by making use of the monotonicity of $\bar u$, we conclude that

$$\bar u(x) = \begin{cases} \beta & -1 \leq x < x_0 \\ \alpha & x_0 < x \leq 1. \end{cases}$$

That the point of discontinuity x_0 is given by the formula (2.4) follows from

$$\int_{-1}^{+1} \bar u dx = \lim_{\varepsilon \to 0}\int_{-1}^{+1} u_\varepsilon dx = M.$$

The proof of part (iii) of the theorem is complete.

3 Proof of proposition

We recall that statement (ii) of the theorem is proved in (Carr, Gurtin and Slemrod 1984) for local minimizers. As before we assume that the family $\{u_\varepsilon\}$ consists of monotonically decreasing functions.

1. $|u_\varepsilon|_{L^\infty} \leq$ Const.
   The argument is as before.

2. $|\sigma_\varepsilon| \leq$ Const.
   The argument is as before.

3. $\int_{-1}^{+1}(u'_\varepsilon)^2 dx \leq \dfrac{\text{Const.}}{\varepsilon^2}$

Multiply the equation (EL) by u_ε and integrate by parts to obtain

$$\varepsilon^2 \int_{-1}^{1} (u'_\varepsilon)^2 dx = -\int_{-1}^{1} W'(u_\varepsilon) u_\varepsilon dx + \sigma_\varepsilon \int_{-1}^{1} u_\varepsilon dx.$$

The estimate now follows from this by utilizing estimates 1 and 2. Note that 3 is a weaker estimate than the corresponding one we had for global minimizers.

4. As before the family $\{u_\varepsilon\}$ is relatively compact by Helly's compactness theorem. Let $\{u_\varepsilon\}$ be a subsequence such that

$$u_\varepsilon \to \bar{u} \quad \text{pointwise on } [-1,1]$$

$$\sigma_\varepsilon \to \bar{\sigma}.$$

Taking the limit in (2.9) by utilizing 3 we obtain

$$W'(\bar{u}) = \bar{\sigma} \text{ a.e. on } [-1,1].$$

This last equation depending on the value of $\bar{\sigma}$ may have one, two or three solutions. Hypothesis (∗) excludes the first two possibilities and so we are left with the case of three roots, $\{r_1, r_2, r_3\}$. We claim that $\bar{u}(x)$ cannot take as a value the middle root r_2 on an open interval I. Indeed if so then by monotonicity the convergence of u_ε to $\bar{u}$ on I is uniform. Computing the second variation of J_ε about a fixed function ϕ in $W^{1,2}$ with compact support in I we find

$$J''_\varepsilon(u_\varepsilon)(\phi,\phi) = \varepsilon^2 \int_{-1}^{+1} (\phi')^2 dx + \int_{-1}^{+1} W''(u_\varepsilon)\phi^2 dx,$$

and utilizing the uniform convergence we conclude that

$$J''_\varepsilon(u_\varepsilon)(\phi,\phi) < 0$$

for ε sufficiently small, a contradiction.

5. To conclude we need to show that $\bar{\sigma} = 0$. For this purpose we argue by contradiction. For definiteness we take $\bar{\sigma} > 0$ and we begin with the case depicted in

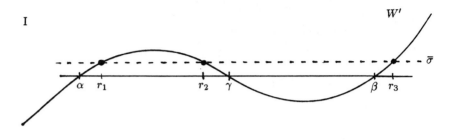

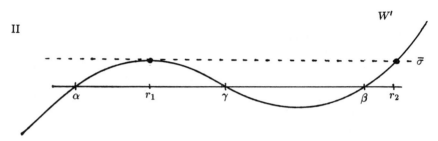

Figure 2. Graph of W' and possible values of $\bar{\sigma}$

Figure 2I below. We also introduce the extra hypothesis on W that besides α and β there is not any other pair of numbers z_1, z_2 such that

$$W'(z_1) = W'(z_2), \quad W(z_1) = W(z_2).$$

By the considerations above

$$\bar{u}(x) = \begin{cases} r_3 & \text{on } [-1, \bar{x}_0] \\ r_1 & \text{on } (\bar{x}_0, 1] \end{cases}$$

for some $\bar{x}_0$ in $(-1, +1)$. Moreover the convergence $u_\varepsilon \to \bar{u}$ is uniform away from $\bar{x}_0$. By the extra hypothesis $W(r_1) \neq W(r_3)$ and so for definiteness we take $W(r_3) < W(r_1)$. In the remaining part of the argument we take ε sufficiently small but fixed and we exhibit elements $\hat{u}_\varepsilon^h$ arbitrarily close to u_ε in $W^{1,2}$ and of mass M with the property that

$$J_\varepsilon(\hat{u}_\varepsilon^h) < J_\varepsilon(u_\varepsilon)$$

thus contradicting that u_ε is a local minimizer in $W^{1,2}$. For $\delta > 0$, small, and arbitrary otherwise set

$$\hat{u}_\varepsilon(x) = \begin{cases} u_\varepsilon(-1+\delta), & x \in [-1, -1+\delta] \\ u_\varepsilon(x), & x \in [-1+\delta, 1]. \end{cases}$$

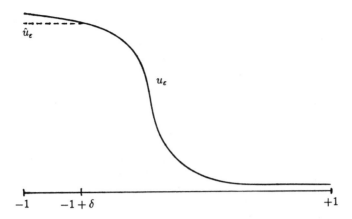

Figure 3. Graphs of the functions u_ε and $\hat{u}_\varepsilon$.

Clearly
$$J_\varepsilon(\hat{u}_\varepsilon) < J_\varepsilon(u_\varepsilon).$$

Next extend $\hat{u}_\varepsilon$ on $(-\infty, -1]$ by setting it there equal to $u_\varepsilon(-1+\delta)$ and define for small $h > 0$ the translated function

$$\hat{u}_\varepsilon^h(x) = \hat{u}_\varepsilon(x - h), \quad x \text{ in } [-1, 1].$$

Clearly
$$\int_{-1}^{+1} \left(\frac{d\hat{u}_\varepsilon^h(x)}{dx}\right)^2 dx \leq \int_{-1}^{+1} \left(\frac{d\hat{u}_\varepsilon(x)}{dx}\right)^2 dx.$$

Also by the extra hypothesis and the assumption that ε is sufficiently small

$$\int_{-1}^{+1} W(\hat{u}_\varepsilon^h(x))dx = \int_{-1}^{+1} W(\hat{u}_\varepsilon(x-h))dx$$
$$= \int_{-1-h}^{1-h} W(\hat{u}_\varepsilon(x))dx = W(\hat{u}_\varepsilon(-1+\delta))h + \int_{-1}^{1-h} W(\hat{u}_\varepsilon(x))dx$$
$$= W(\hat{u}_\varepsilon(-1+\delta))h + \int_{-1}^{+1} W(\hat{u}_\varepsilon(x))dx - \int_{1-h}^{1} W(\hat{u}_\varepsilon(x))dx$$
$$\leq \int_{-1}^{+1} W(\hat{u}_\varepsilon(x))dx,$$

and therefore
$$J_\varepsilon(\hat{u}_\varepsilon^h) \leq J_\varepsilon(\hat{u}_\varepsilon).$$

Now by cutting off the function u_ε and producing $\hat{u}_\varepsilon$ we decreased slightly its mass. On the other hand by translating $\hat{u}_\varepsilon$ to the right we can obviously gain the lost mass and moreover the element $\hat{u}_\varepsilon^h$ can be chosen as close (in $W^{1,2}$) to u_ε as desired. This establishes the sought contradiction in the case depicted in Figure 2I. The argument for handling Figure 2II is similar.

We established that $\bar{\sigma} = 0$ under the extra hypothesis above. We now sketch a different argument, due to Peter Bates, that establishes that $\bar{\sigma} = 0$ without the extra hypothesis and under the sole requirement that the total variations of the u_ε's are uniformly bounded as $\varepsilon \to 0$. Again we argue by contradiction and so take $\bar{\sigma} \neq 0$. We then observe that the equation

$$\ddot{U} = W'(U) - \bar{\sigma}$$

has only homoclinic orbits. From the uniform bound on the variations, oscillations are excluded and by an argument involving stretching of variables we conclude that in the stretched variables u_ε converges to a homoclinic orbit. It follows that in the original variables u_ε converges to a constant. However this is excluded by hypothesis (∗), and so the result follows once more.

REMARK: The estimates obtained in §4 hold for this class of local minimizers as can be seen from the proofs that do not assume more than that $\{u_\varepsilon\}$ converges to the single-interface solution and that $\sigma_\varepsilon \to 0$ as $\varepsilon \to 0$.

4   Proof of estimate (1.3)

We begin with some definitions, and some remarks that follow from part (iii) above. Set

$$\beta_\varepsilon = u_\varepsilon(-1), \quad \alpha_\varepsilon = u_\varepsilon(1).$$

Then we have that

(4.1) $$\beta_\varepsilon \to \beta, \quad \alpha_\varepsilon \to \alpha \text{ as } \varepsilon \to 0.$$

Let

(4.2) $$M_u = \max(-u_\varepsilon''(-1), \; u_\varepsilon''(+1)),$$

and recall (2.1)

(4.3) $$-u_\varepsilon''(-1) \geq 0, \; u_\varepsilon''(+1) \geq 0.$$

Choose $\delta > 0$ so that

(4.4) $$\alpha < \alpha + \delta < P_1 < P_2 < \beta - \delta < \beta$$

and arbitrary otherwise and keep it fixed. Define x_β, x_α as the (unique) roots of the equations

$$u_\varepsilon(x) = \beta - \delta, \; u_\varepsilon(x) = \alpha + \delta.$$

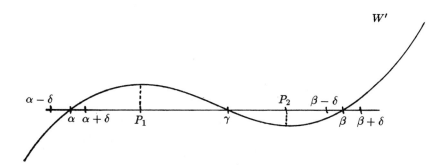

Figure 4. Graph of W' showing location of P_1 and P_2.

By (iii)

(4.5) $$x_\beta, x_\alpha \to x_0 \text{ as } \varepsilon \to 0.$$

Let

(4.6) $$q^2 = q^2(\delta) := \min\{W''(s) | s \in [\alpha - \delta, \alpha + \delta] \cup [\beta - \delta, \beta + \delta]\},$$

$$c_1 = \frac{\beta - \beta_\varepsilon}{\beta_\varepsilon - \alpha_\varepsilon}, \quad c_2 = \frac{\alpha_\varepsilon - \alpha}{\beta_\varepsilon - \alpha_\varepsilon}.$$

1. The following estimates hold true:

(4.7) $$-c_2 W'(\alpha_\varepsilon) \leq \sigma_\varepsilon \leq -c_1 W'(\beta_\varepsilon)$$

(4.8) $$c_1 \geq 0, \ c_2 \geq 0$$

(In terms of u_ε, $\beta \geq \beta_\varepsilon = u_\varepsilon(-1) > u_\varepsilon(1) = \alpha_\varepsilon \geq \alpha$).

PROOF OF (4.7):
On $(-1, 1)$ u_ε satisfies

(4.9) $$\varepsilon^2 u_\varepsilon'' = W'(u_\varepsilon) - \sigma_\varepsilon$$

and on the boundary

(4.10) $$u_\varepsilon'(-1) = u_\varepsilon'(1) = 0.$$

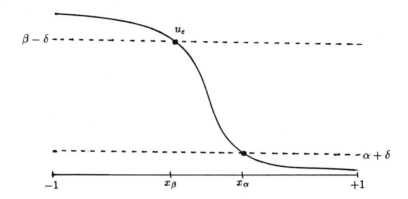

Figure 5. Graph of u_ε showing location of x_α and x_β.

Multiplying equation (4.9) by u'_ε and integrating over $[-1,1]$ we obtain

$$\frac{\varepsilon^2}{2}(u'_\varepsilon(x))^2|^1_{-1} = \int_{\beta_\varepsilon}^{\alpha_\varepsilon} W'(s)ds - \sigma_\varepsilon u_\varepsilon(x)|^{+1}_{-1},$$

and so using the boundary conditions we find

$$\begin{aligned}\sigma_\varepsilon(\beta_\varepsilon - \alpha_\varepsilon) &= \int_{\alpha_\varepsilon}^{\beta_\varepsilon} W'(s)ds \\ &= \int_{\alpha_\varepsilon}^{\alpha} W'(s)ds + \int_{\alpha}^{\beta} W'(s)ds + \int_{\beta}^{\beta_\varepsilon} W'(s)ds \\ &= -\int_{\alpha}^{\alpha_\varepsilon} W'(s)ds + \int_{\beta}^{\beta_\varepsilon} W'(s)ds.\end{aligned}$$

Now, for ε sufficiently small, $|\alpha - \alpha_\varepsilon| < \delta$ with δ as in (4.4), and therefore (see Figure 4) we obtain

(4.11) $$\int_{\alpha}^{\alpha_\varepsilon} W'(s)ds \geq 0, \quad \int_{\beta}^{\beta_\varepsilon} W'(s)ds \geq 0.$$

Using these facts we conclude from above that

(4.12) $$-\int_{\alpha}^{\alpha_\varepsilon} W'(s)ds \leq \sigma_\varepsilon(\beta_\varepsilon - \alpha_\varepsilon) \leq \int_{\beta}^{\beta_\varepsilon} W'(s)ds.$$

Using the monotonicity of W' in $(\alpha - \delta, \alpha + \delta) \cup (\beta - \delta, \beta + \delta)$ we obtain

(4.13) $$\int_{\alpha}^{\alpha_\varepsilon} W'(s)ds \leq (\alpha_\varepsilon - \alpha)W'(\alpha_\varepsilon)$$

(4.14) $$\int_{\beta}^{\beta_\varepsilon} W'(s)\,ds \leq (\beta_\varepsilon - \beta)W'(\beta_\varepsilon),$$

and making use of (4.13),(4.14), and (4.12) we obtain (4.7).

PROOF OF (4.8):
We proceed by contradiction. So assume that $c_1 < 0$. Thus $\beta_\varepsilon > \beta$. From (4.9),(4.3) we have

(4.15) $$0 \geq W'(\beta_\varepsilon) - \sigma_\varepsilon$$

and so

(4.16) $$(-c_1)\sigma_\varepsilon \geq (-c_1)W'(\beta_\varepsilon).$$

From $\beta_\varepsilon > \beta$ and (4.1) we obtain that for ε small $W'(\beta_\varepsilon) > 0$ and so by (4.15) $\sigma_\varepsilon > 0$. On the other hand by (4.7)

(4.17) $$0 < \sigma_\varepsilon \leq (-c_1)W''(\beta_\varepsilon).$$

Combining (4.16) and (4.17) we obtain

(4.18) $$0 < \sigma_\varepsilon \leq (-c_1)\sigma_\varepsilon$$

from which it follows that

(4.19) $$-c_1 \geq 1 \Leftrightarrow \alpha_\varepsilon \geq \beta,$$

which by (4.1) is in contradiction to $\beta > \alpha$. This argument establishes half of (4.8). The rest can be established by analogous reasoning. We omit the details.

2. Define $\bar{c} = \max(c_1, c_2)$. Then for ε sufficiently small we have the estimates

(4.20) $$W'(\alpha_\varepsilon) \geq 0, \quad W'(\beta_\varepsilon) \leq 0$$

(4.21) $$\bar{c}(-W'(\alpha_\varepsilon)) \leq \sigma_\varepsilon \leq \bar{c}(-W'(\beta_\varepsilon))$$

(4.22) $$|\sigma_\varepsilon| \leq \bar{c}M_u\varepsilon^2$$

(4.23) $$|W'(\alpha_\varepsilon)| \leq 2\varepsilon^2 M_u$$

(4.24) $$|W'(\beta_\varepsilon)| \leq 2\varepsilon^2 M_u$$

(4.25) $$|\alpha - \alpha_\varepsilon| \leq \frac{2\varepsilon^2 M_u}{q^2}$$

(4.26) $$|\beta - \beta_\varepsilon| \leq \frac{2\varepsilon^2 M_u}{q^2}.$$

PROOF OF (4.20)–(4.26):

(4.20), (4.21) follow from (4.7), (4.8). To see (4.22) evaluate (4.9) at ± 1 to obtain
$$\varepsilon^2 u_\varepsilon''(-1) = W'(\beta_\varepsilon) - \sigma_\varepsilon$$
$$\varepsilon^2 u_\varepsilon''(+1) = W'(\alpha_\varepsilon) - \sigma_\varepsilon.$$

Then,

(4.27) $$-W'(\beta_\varepsilon) = -\sigma_\varepsilon - \varepsilon^2 u_\varepsilon''(-1)$$
(4.28) $$-W'(\alpha_\varepsilon) = -\sigma_\varepsilon - \varepsilon^2 u_\varepsilon''(+1).$$

If $\sigma_\varepsilon > 0$
$$-W'(\beta_\varepsilon) \leq -\varepsilon^2 u_\varepsilon''(-1) \leq M_u \varepsilon^2,$$

and if $\sigma_\varepsilon < 0$
$$-W'(\alpha_\varepsilon) \geq -\varepsilon^2 u_\varepsilon''(+1) \geq -M_u \varepsilon^2.$$

(4.22) then follows from (4.20), (4.21). (4.23) follows from (4.28), (4.22) and the observation that $\bar{c}$ can be taken without loss of generality less than 1. (4.24) follows in the same way from (4.27). Finally to see (4.26) we write

(4.29)
$$\begin{aligned} W''(\xi)(\beta - \beta_\varepsilon) &= W'(\beta) - W'(\beta_\varepsilon) \\ &= -W'(\beta_\varepsilon) \end{aligned}$$

where ξ is some point in $(\beta_\varepsilon, \beta)$ and therefore by (4.1) and (4.6), $W''(\xi) \geq q^2$ and so (4.26) follows from (4.29) and (4.24). The proof of (4.25) is similar. Combining (4.25), (4.26) with (4.22) we obtain

(4.30) $$|\sigma_\varepsilon| \leq \frac{2\varepsilon^4 M_u^2}{q^2(\beta_\varepsilon - \alpha_\varepsilon)}.$$

3. Next we obtain a refinement of the result in part (iii) of the theorem above.
The following estimates hold true for ε sufficiently small

(4.31) $$0 \leq \beta_\varepsilon - u_\varepsilon(x) \leq \text{(Const.)} \, e^{-\frac{q}{\varepsilon}(x_\beta - x)} \text{ for } -1 \leq x \leq x_\beta$$

(4.32) $$0 \leq u_\varepsilon(x) - \alpha_\varepsilon \leq \text{(Const.)} \, e^{-\frac{q}{\varepsilon}(x - x_\alpha)} \text{ for } x_\alpha \leq x \leq 1.$$

We establish (4.31). The derivation of (4.32) is analogous. For $x \in [-1, x_\beta]$
$$\varepsilon^2 u_\varepsilon'' = W'(u_\varepsilon) - \sigma_\varepsilon$$

and therefore
$$\varepsilon^2 (u_\varepsilon - \beta_\varepsilon)'' - (W'(u_\varepsilon) - W'(\beta_\varepsilon)) = W'(\beta_\varepsilon) - \sigma_\varepsilon = \varepsilon^2 u_\varepsilon''(-1) \leq 0.$$

Hence, by the intermediate value theorem

$$\varepsilon^2(\beta_\varepsilon - u_\varepsilon)'' - W''(\xi)(\beta_\varepsilon - u_\varepsilon) \geq 0$$

and so

$$\varepsilon^2 v'' - q^2 v \geq (W''(\xi) - q^2)v$$

where

$$v = \beta_\varepsilon - u_\varepsilon \geq 0.$$

Now for x in $[-1, x_\beta]$, u_ε takes values in $[\beta - \delta, \beta_\varepsilon]$ and therefore by taking ε sufficiently small we can assume that $W''(\xi) \geq q^2$. So

(4.33) $$\varepsilon^2 v'' - q^2 v \geq 0$$

(4.34) $$v'(-1) = 0, \quad v(x_\beta) = \beta_\varepsilon + \delta - \beta.$$

Next we compare v with the solution of

(4.35) $$\varepsilon^2 y'' - q^2 y = 0$$
$$y'(-1) = 0, \quad y(x_\beta) = \beta_\varepsilon - \beta + \delta,$$

which can be computed explicitly to be

(4.36) $$y(x) = (\beta_\varepsilon - \beta + \delta)\cosh^{-1}\left(\frac{q}{\varepsilon}(x_\beta + 1)\right)\cosh\left(\frac{q}{\varepsilon}(x + 1)\right).$$

The maximum principle yields

(4.37) $$0 \leq v(x) \leq y(x).$$

Moreover noting that for ε sufficiently small (by (4.8))

$$\beta_\varepsilon - \beta + \delta \leq \delta$$

and by making use of

$$\cosh^{-1}\left(\frac{q}{\varepsilon}(x_\beta + 1)\right) \leq 2\exp\left(-\frac{q}{\varepsilon}(x_\beta + 1)\right)$$

we obtain from (4.36) that

(4.38) $$y(x) \leq 2\delta e^{-\frac{q}{\varepsilon}(x_\beta - x)}$$

and so (4.31) follows.

4. Let

$$M_w^2 = \max\{W'''(s) | s \in [\alpha - \delta, \alpha + \delta] \cup [\beta - \delta, \beta + \delta]\}.$$

Clearly then on $[-1, x_\beta] \cup [x_\alpha, 1]$ we have

(4.39) $$q^2 \leq W''(u_\varepsilon(x)) \leq M_w^2.$$

The following estimates hold true:

(4.40) $$0 \leq -u_\varepsilon'(x) \leq -u_\varepsilon''(-1)\varepsilon + \frac{(M_w + 1)(\beta_\varepsilon - u_\varepsilon(x))}{\varepsilon}, \quad x \in [-1, x_\beta]$$

(4.41) $$0 \leq -u_\varepsilon'(x) \leq -u_\varepsilon''(+1)\varepsilon + \frac{(M_w + 1)(u_\varepsilon(x) - \alpha_\varepsilon)}{\varepsilon}, \quad x \in [x_\alpha, 1].$$

PROOF OF (4.40):

Let $x \in [-1, x_\beta]$. Multiplying (4.9) by u_ε' and integrating over $[-1, x]$ we obtain

(4.42) $$\begin{aligned}\frac{\varepsilon^2}{2}(u_\varepsilon'(x))^2 &= \int_{\beta_\varepsilon}^{u_\varepsilon(x)}(W'(s) - \sigma_\varepsilon)ds \\ &= \sigma_\varepsilon(\beta_\varepsilon - u_\varepsilon(x)) + W(u_\varepsilon(x)) - W(\beta_\varepsilon).\end{aligned}$$

Using Taylor's theorem we find

$$W(u_\varepsilon) = W(\beta_\varepsilon) + (u_\varepsilon - \beta_\varepsilon)W'(\beta_\varepsilon) + \frac{(u_\varepsilon - \beta_\varepsilon)^2}{2}W''(\xi)$$

where ξ is a point in $[\beta - \delta, \beta_\varepsilon]$. Therefore $W''(\xi) \leq M_w^2$ and so by (4.42)

(4.43) $$\frac{\varepsilon^2}{2}(u_\varepsilon'(x))^2 \leq (\beta_\varepsilon - u_\varepsilon(x))(\sigma_\varepsilon - W'(\beta_\varepsilon)) + \frac{M_w^2}{2}(\beta_\varepsilon - u_\varepsilon(x))^2.$$

Using that

$$\varepsilon^2(-u_\varepsilon''(-1)) = \sigma_\varepsilon - W'(\beta_\varepsilon)$$

we have

(4.44) $$\frac{\varepsilon^2}{2}(u_\varepsilon'(x))^2 \leq (\beta_\varepsilon - u_\varepsilon(x))\varepsilon^2(-u_\varepsilon''(-1)) + \frac{M_w^2}{2}(\beta_\varepsilon - u_\varepsilon(x))^2$$

and so

$$\begin{aligned}(u_\varepsilon'(x))^2 &\leq \frac{(\beta_\varepsilon - u_\varepsilon(x))^2}{\varepsilon^2} + \varepsilon^2(-u_\varepsilon''(-1))^2 + \frac{M_w^2}{\varepsilon^2}(\beta_\varepsilon - u_\varepsilon(x))^2 \\ &\leq \varepsilon^2(-u_\varepsilon''(-1))^2 + \frac{(\beta_\varepsilon - u_\varepsilon(x))^2}{\varepsilon^2}(M_w^2 + 1),\end{aligned}$$

from which we obtain

(4.45) $$0 \leq -u_\varepsilon'(x) \leq \varepsilon(-u_\varepsilon''(-1)) + \frac{\beta_\varepsilon - u_\varepsilon(x)}{\varepsilon}(M_w + 1).$$

(4.41) is established similarly.

5. *The following estimates hold true:*

(4.46) $$-u_\varepsilon''(-1) \le \frac{\text{Const.}}{\varepsilon^2} e^{-\frac{q}{\varepsilon}(x_\beta + 1)}$$

(4.47) $$u_\varepsilon''(+1) \le \frac{\text{Const.}}{\varepsilon^2} e^{-\frac{q}{\varepsilon}(1 - x_\alpha)}.$$

PROOF OF (4.46):
Let x in $[-1, x_\beta]$. u_ε is decreasing and therefore so is $W'(u_\varepsilon)$. From the equation
$$\varepsilon^2 u_\varepsilon'' = W'(u_\varepsilon) - \sigma_\varepsilon$$
we conclude that u_ε'' is decreasing. Therefore
$$0 \le -u_\varepsilon''(-1) \le \frac{1}{1+x} \int_{-1}^{x} -u_\varepsilon''(s)\,ds = -\frac{u_\varepsilon'(x)}{1+x},$$
and so

(4.48) $$0 \le -u_\varepsilon''(-1) \le -\frac{u_\varepsilon'(x)}{1+x}.$$

Therefore by (4.44)

(4.49) $$(-u_\varepsilon''(-1))^2 (1+x)^2 \le (-u_\varepsilon(x))^2$$
$$\le 2(\beta_\varepsilon - u_\varepsilon(x))(-u_\varepsilon''(-1)) + \frac{M_w^2}{\varepsilon^2}(\beta_\varepsilon - u_\varepsilon(x))^2.$$

Take now $x = -1 + \frac{\varepsilon}{q}$, an acceptable choice for ε sufficiently small by (4.5), to find that
$$(-u_\varepsilon''(-1))^2 \left(\frac{\varepsilon}{q}\right)^2 \le 2(\beta_\varepsilon - u_\varepsilon(-1 + \frac{\varepsilon}{q}))(-u_\varepsilon''(-1))$$
$$+ \frac{M_w^2}{\varepsilon^2}(\beta_\varepsilon - u_\varepsilon(-1 + \frac{\varepsilon}{q}))^2.$$

Solving this inequality we obtain

(4.50) $$0 \le -u_\varepsilon''(-1) \le \frac{2q^2 + qM_w}{\varepsilon^2}(\beta_\varepsilon - u_\varepsilon(-1 + \frac{\varepsilon}{q})).$$

By (4.31)
$$0 \le \beta_\varepsilon - u_\varepsilon(-1 + \frac{\varepsilon}{q}) \le (\text{Const.}) e^{-\frac{q}{\varepsilon}(x_\beta + 1)}$$
and so (4.46) follows. The derivation of (4.47) is similar.

6. Let d be the distance of x_0 from the boundary of the interval $[-1,1]$ and ν^2 a number smaller than $\min\{W''(\alpha), W''(\beta)\}$ but as close to it as desired. Then the following estimate holds true for ε sufficiently small:

$$(4.51) \qquad |\sigma_\varepsilon| \leq (\text{Const.}) e^{-\frac{2\nu}{\varepsilon}d}.$$

The proof of (4.51) follows from (4.46), (4.47), (4.30) and (4.5). The proof of estimate (1.3) is complete.

Notes

(i) Making use of (4.31), (4.32) in (4.40), (4.41) we obtain

$$(4.52) \qquad 0 \leq -u'_\varepsilon(x) \leq -u''_\varepsilon(-1)\varepsilon + \frac{(\text{Const.})}{\varepsilon} e^{-\frac{q}{\varepsilon}(x_\beta - x)}, \quad x \in [-1, x_\beta]$$

$$(4.53) \qquad 0 \leq -u'_\varepsilon(x) \leq u''_\varepsilon(+1)\varepsilon + \frac{(\text{Const.})}{\varepsilon} e^{-\frac{q}{\varepsilon}(x - x_\alpha)}, \quad x \in [x_\alpha, 1].$$

By (4.46) and (4.47) we see that in the estimates above the second terms are the determining factors, and therefore we have

$$(4.54) \qquad 0 \leq -u'_\varepsilon(x) \leq \frac{(\text{Const.})}{\varepsilon} e^{-\frac{q}{\varepsilon}(x_\beta - x)}, \quad x \in [-1, x_\beta]$$

$$(4.55) \qquad 0 \leq -u'_\varepsilon(x) \leq \frac{(\text{Const.})}{\varepsilon} e^{-\frac{q}{\varepsilon}(x - x_\alpha)}, \quad x \in [x_\alpha, 1].$$

From these estimates we can obtain corresponding statements for u''_ε. Indeed first let x be in $[-1, x_\beta]$. By differentiating (4.9) we obtain

$$(4.56) \qquad \varepsilon^2 u'''_\varepsilon = W''(u_\varepsilon) u'_\varepsilon$$

and therefore by substituting in

$$\begin{aligned}
-u''_\varepsilon(x) &= -u''_\varepsilon(-1) + \int_{-1}^x -u'''_\varepsilon(s)\,ds \\
&= -u''_\varepsilon(-1) + \int_{-1}^x \frac{W''(u_\varepsilon)}{\varepsilon^2}(-u'_\varepsilon(s))\,ds \\
&\leq \frac{(\text{Const.})}{\varepsilon^2} e^{-\frac{q}{\varepsilon}(x_\beta + 1)} + \frac{(\text{Const.})}{\varepsilon^2} \int_{-1}^x (-u'_\varepsilon(s))\,ds
\end{aligned}$$

(by (4.46))

$$\begin{aligned}
&\leq \frac{(\text{Const.})}{\varepsilon^2} e^{-\frac{q}{\varepsilon}(x_\beta + 1)} + \frac{(\text{Const.})}{\varepsilon^2}(\beta_\varepsilon - u_\varepsilon(x)) \\
&\leq \frac{(\text{Const.})}{\varepsilon^2} e^{-\frac{q}{\varepsilon}(x_\beta + 1)} + \frac{(\text{Const.})}{\varepsilon^2} e^{-\frac{q}{\varepsilon}(x_\beta - x)}
\end{aligned}$$

(by (4.31))
$$\leq \frac{(\text{Const.})}{\varepsilon^2} e^{-\frac{q}{\varepsilon}(x_\beta - x)}$$
(since x is in $[-1, x_\beta]$). Therefore we obtain

(4.57) $$0 \leq -u_\varepsilon''(x) \leq \frac{(\text{Const.})}{\varepsilon^2} e^{-\frac{q}{\varepsilon}(x_\beta - x)}, \quad x \in [-1, x_\beta]$$

and similarly

(4.58) $$0 \leq -u_\varepsilon''(x) \leq \frac{(\text{Const.})}{\varepsilon^2} e^{-\frac{q}{\varepsilon}(x - x_\alpha)}, \quad x \in [x_\alpha, 1].$$

(ii) Using estimate (2.7) that holds for global minimizers we can establish easily the estimate

(4.59) $$|x_\beta - x_\alpha| \leq (\text{Const.})\varepsilon$$

which is a refinement over (4.5) and holds for global minimizers.

References

[1] Alikakos, N. D. and Shaing, K. C. (1987). On the singular limit for a class of problems modeling phase transitions. *SIAM Journal on Mathematical Analysis*, **18**, 1453-62.

[2] Cahn, J. W.(1961). On spinodal decomposition. *Acta Metallurgica*, **9**, 795-801.

[3] Cahn, J. W. and Hilliard, J. E. (1958). Free energy of a nonuniform system I, interfacial free energy. *Journal of Chemical Physics*, **28**, 258-67.

[4] Carr, J., Gurtin, M. and Slemrod, M. (1984). Structural phase transitions on a finite interval. *Archive for Rational Mechanics and Analysis*, **86**, 317-51.

[5] Gurtin, M. E. (1985). *Some results and conjectures in the gradient theory of phase transitions.* Technical Report **156**, Institute of Mathematics and its Applications, University of Minnesota.

[6] Luckhaus, S. and Modica, L., The Gibbs-Thompson relation within the gradient theory of phase transitions. Preprint.

[7] Natanson, I. P. (1961). *Theory of functions of a real variable*, Vol I, English Translation. Ungar.

[8] Van der Waals, J. D.(1893). The thermodynamic theory of capillarity under the hypothesis of a continuous variation of density, (in Dutch). *Koninklÿke Akademie voor Wetenschnappen Amsterdam (section 1)*, **1**, No. 8.

Excitability behaviour of myelinated axon models

Jonathan Bell [1]
University at Buffalo, SUNY

1 Introduction

One of the more quantitative areas of biology is celluar neurobiology. The reason for this is that there are relatively well developed experimental techniques and good experimental data which are necessary for the development of models. Also, we can rely theoretically on the conservation principles. A particular subarea of this field is axonology where we consider the excitability behaviour of axonal processes associated with single nerve cells under various stimulation conditions. An important, yet specialized, such process is the myelinated axon in higher animals. A goal of this paper is to review analytical results on some simple models of myelinated axons. The models take the form of either reaction-diffusion equations or differential-difference equations coupled to ordinary differential equations whose variables represent subsidiary activation processes in the excitable membrane. The principle variable is always the transmembrane potential, and all these systems are excitable in the sense that there is a locally stable rest state for the system, but if a perturbation of sufficient magnitude is imposed on the problem, then the potential takes a large excursion from the rest state. Depending on the model, it can either go to another equilibrium value or return to its original rest state as $t \to \infty$.

In the next section we review some biology background to understand the type of problems discussed in the rest of the paper. In the third section we discuss recent work on diffusion models, and in the fourth section we impose slightly different modelling assumptions to arrive at a spatially discrete class of models. Then we discuss the stability of their rest state.

2 Background

In cellular neurobiology we are concerned with the behaviour of the voltage changes across the biologically excitable membrane. A particular part of the nerve cell of interest here is the long axonal process extending from the cell body which can be thought of as a leaky cable. Under normal rest conditions there are different concentrations of various ions on both sides of the axon's membrane, and this sets up a rest potential. The most prominent ionic species are sodium, potassium, and

---

[1] Partially supported by NSF grant DMS-8615739. This paper was begun while the author was a visitor at the Department of Mathematics, Heriot-Watt University, Edinburgh.

calcium, and a typical rest potential is $E_R \cong -70\text{mV}$. The primary variable of interest is the transmembrane potential, E, which is the difference between inside and outside potentials. In this paper all potential values are taken with respect to the rest potential, so our dynamic variable of interest is $v = E - E_R$. If one considers a small section of the axon, insulates its ends and injects a current into the inside, the ionic concentration balance across the membrane is changed, and hence the membrane potential is altered. For small current injections of short duration the potential eventually returns to its rest potential, here $v = 0$. This is considered a subthreshold response and such experiments are useful in determining certain properties of the membrane. If the current stimulus is sufficiently strong, the potential takes an excursion on the order of 100 mV before returning to rest (Figure 1). This superthreshold response is fairly robust to the type of initiating stimulus. In the membrane is a distribution of ionic channels, each rather selective to a particular ion. Each channel also, classically, has a giant mechanism which is voltage-sensitive. When a threshold stimulus is applied, the potential is changed, affecting the rate of opening of these channels. Various ionic channels work on different time scales, but classically we can illustrate the case by discussing the giant axon of squid Loligo. First sodium channels are gated open when potential is large enough, allowing a sodium current to flow into the cell. As a feedback mechanism this tends to depolarize the cell, i.e., drive v more positive. On a slower time scale potassium channels are opened when the potential is large, and the effect of the outward potassium current tends to drive the potential back to rest.

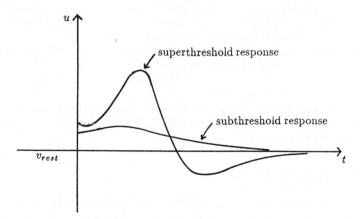

Figure 1. Possible potential variations.

The membrane current density is considered a sum of two components, a capacitance current and a resistive, or ionic current density I_{ion}. Thus the membrane current density is

(2.1) $$I_m = C_m dv/dt + I_{ion} ,$$

where C_m is a membrane capacity, typically 1 $\mu F/cm^2$. The simplest expression for I_{ion} is to write
$$I_{ion} = \sum_j g_j(v - v_j)$$
where the sum is taken over all ionic species associated with the particular membrane, with g_j being the (voltage-sensitive) conductance, and v_j being the (constant) reversal potential of the jth species. In the seminal work of Hodgkin and Huxley(1952), they wrote I_{ion} as a sum of sodium, potassium, and residual (or leakage) currents:

(2.2) $$I_{ion} = g_{Na}(v - v_{Na}) + g_k(v - v_K) + g_\ell(v - v_\ell).$$

Although g_ℓ is a constant estimated from experiment and hence the leakage current is ohmic, the sodium and potassium conductances depend on voltage in a complicated way(see Tuckwell(1988), or Hodgkin and Huxley(1952) for details). We shall not further describe the conductances here, but only say that each conductance depends on inactivation and/or activation variables whose dynamics depend on voltage which is specified by curve fitting voltage-clamp experimental data.

The axon is really a cable and as such propagates the voltage impulse discussed above. Since the length of the axon is much longer than the axonal radius, the axoplasm at the cross-section is considered isopotential, so circuit models for the axon are almost always spatially one-dimensional, which takes into account at each x the local dynamics (2.1). Therefore, conservation of current hypothesis on the circuit leads to the cable model

(2.3) $$C_m \partial v/\partial t + I_{ion}(v) = (a/2R_i)\partial^2 v/\partial x^2,$$

if scaled with respect to the ratio of radius, a, and to axoplasmic resistivity R_i. Thus, a propagating impulse, or action potential, is a travelling wave solution $v(x,t) = V(x - \theta t)$ to the equation (2.3).

In the case of the Hodgkin-Huxley theory, $I_{ion}(v)$ is really a function of three other activation/inactivation processes, each of which satisfy an equation of the form
$$\partial y/\partial t = [y_\infty(v) - y]/\tau(v).$$

Since major theoretical questions concern existence and stability of action potentials, or trains of such potentials, as well as how such action potentials are initiated, (2.3) coupled to the system of three ordinary differential equations is a complicated system to handle analytically. Some progress has been made under certain assumptions, but various approaches have concentrated on looking at caricatures of the Hodgkin-Huxley model. One such caricature is the FitzHugh-Nagumo model, which may be written as

$$v_t - f(v) + w = v_{xx}, \quad w_t = \sigma v - \gamma w,$$

where we have dropped nonessential constants, and have gone over to a subscript notation for partial derivatives. Variable v has the same meaning as before, while w is a "recovery" variable representing the dynamics of all slow processes (like potassium conductance). Here σ and γ are nonnegative rate constants, and the qualitative nature of the current-voltage relation $f(v)$ is shown in Figure 2. Much work has been done on this model regarding travelling wave solutions, as well as periodic wave trains, and their stability, but our purpose for introducing it here is to use the membrane dynamics description in our model of myelinated axon nodes.

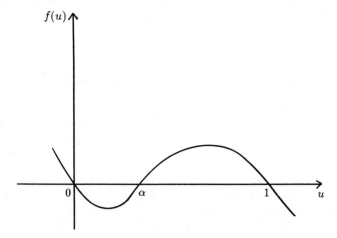

Figure 2. Current-voltage relation.

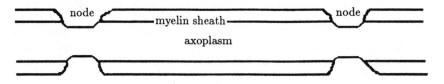

Figure 3. Myelinated axon.

All this model description is with respect to a uniform, unmyelinated axon, which means every section of the axon is equivalent to every other section. Mammals have a large percentage of their axonal processes myelinated, which means the axon are wrapped in a lipoprotein sheath called myelin, which insulates most of the axon. A schematic of a myelinated axon is given in Figure 3. Fibres in the peripheral nervous system have myelin segments roughly 1 to 2 millimetres long, separated by gaps, called nodes of Ranvier, which are a few micrometres wide (Waxman 1972). The nodes provides a conducting path of extracellular fluid to the excitable membrane of the node. Stimulation at one node generates longitudinal ionic currents which excite neighbouring nodes. Because of the insulating nature of the myelinated segments, impulses have the character of jumping from node to node. This form of "saltatory" conduction evolved because only a small percentage of membrane needs to be excited for propagating impulses. Hence conduction takes much less energy per unit length than in unmyelinated axons, and with a much higher conduction speed for a given radius (above some critical minimum radius).

Because of the high resistance and low capacitance of myelin, and also because of the low density of ionic channels found in the internode, the potential across the internode is considered ohmic in behaviour. FitzHugh (1962) gave a model based on the circuit model given in Figure 4, where the nodes are so short they are taken as point sources of excitation (see also Goldman and Albus(1968)). The model using conservation of current principles becomes (with nodes at $x = x_n$, $n \in \mathbf{Z}$)

(2.4)
$$C_m V_t + v/r = (a/2R_i)v_{xx} \qquad x \neq x_n$$
$$C_m dv_n/dt + I_{ion}(v_n) = R^{-1}[[v_x]]_{x_n} \qquad x = x_n$$

where $v_n(t) = \lim_{x \to x_n} v(x,t)$ is the membrane potential at the nth node, and $[[p]]_x$ means $\lim_{\varepsilon \to 0+}[p(x+\varepsilon) - p(x-\varepsilon)]$.

In the next section we consider a scaled form of (2.4), where we impose FitzHugh-Nagumo dynamics at the nodes, then examine some questions associated with excitability of the model. In the section we also introduce briefly another diffusion modelling approach where the nodes do have length. It is the case that in the central nervous system the nodal area can occupy up to twenty percent of the surface area of the axon; that is, it is certainly not reasonable in that case to consider point nodes.

3 Diffusion models

We first consider the FitzHugh formulation of a model for myelinated axon, namely (2.4). If we rescale the internodal length to 1 and have the nodes at integer values

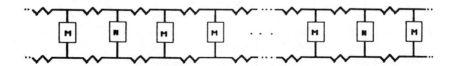

Figure 4. FitzHugh circuit model

of x, then the first model to be considered takes the form

(3.1) $$\begin{cases} Cv_t + gv = v_{xx} & x \in \mathbf{R}\setminus\mathbf{Z} \\ \left. \begin{array}{l} v_t - f(v) + w = \beta[[v_x]]_n \\ w_t = \sigma v - \gamma w \end{array} \right\} & x = n \in \mathbf{Z} \end{cases}$$

To set up as an initial value problem (IVP) we would also impose an initial distribution of potential; that is,

(3.2) $$v(x,0) = \varphi(x), \; w(n,0) = h_n \; x \in \mathbf{R}, \; n \in \mathbf{Z},$$

where φ is some specified continuous function. An experimentally more realistic situation is to consider the axon initially at rest and stimulate it at one node, say at $n = 0$. Then we might just consider a semi-infinite axon and impose

(3.3) $$\begin{aligned} v(x,0) &= w(x,0) = 0 & x > 0 \\ v_x(0,t) &= -I_{app}(t) & t > 0. \end{aligned}$$

Here $I_{app}(\cdot) > 0$ is a specified current (from an electrode). In this paper we will just consider the IVP on an infinite axon, since more work has been conducted on this problem. For some initial work on the initial-boundary value problem, see Bell and Cosner (1986). In this subsection we write the initial conditions as (φ, ψ, h), where $h = (h_n)$, and $\psi = (\varphi_n) = (\varphi(n,0))$.

For purposes of the existence theory, let $C = \beta = 1$, and let g, σ, γ be fixed constants. The approach can allow for considerable generalization of the dynamics, but this was not taken because we wanted to explore qualitative behaviour of a model with dynamics familiar to biomathematicans. We consider the existence of a solution $(v(\cdot,t),(v_j(t)),w_j(t))$, for $t \in [0,T]$, where $v(j,t) = v_j(t)$ for all $t \in [0,T]$, all $j \in \mathbf{Z}$.

We first review the existence theory for (3.1), which appears in detail in Chen and Bell(1988). There we assume (3.2) on $\mathbf{R}$, with φ continuous and bounded on $\mathbf{R}$, f is extendable to a function on $\mathbf{C}$, which viewed as a function of two variables, is C^1, and whose first derivative is bounded. Then the problem is written in an abstract form so that the existence machinery in Friedman (1969) can be applied. To outline this approach, define

$$H \equiv \{(p,q,r) \mid p \in L^2(\mathbf{R}\backslash\mathbf{Z}), q, r \in \ell^2(\mathbf{R})\}, \text{ with}$$

inner product

$$\langle(p_1,q_1,r_1),(p_2,q_2,r_2)\rangle = \int_{\mathbf{R}\backslash\mathbf{Z}} p_1 \bar{p}_2 dx + \sum_i (q_1^i \bar{q}_2^i + r_1^i \bar{r}_2^i),$$

and

$$X \equiv \{(p,q,r) \in H \mid p', p'' \in L^2(\mathbf{R}\backslash\mathbf{Z}), p' \text{ is absolutely continuous}$$

$$\text{on } (j-1,j) \text{ for each } j \in \mathbf{Z}, ([[p']]_j)_{\mathbf{Z}} \in \ell^2, \lim_{x \to j} p(x) = q^j\}.$$

H is a complete complex Hilbert space, and X is dense in H. We define a linear operator $A: X \to H$ by

$$A(u,v,w) \equiv (-u_{xx} + u, ([[-u_x]]_j + v^j)_{\mathbf{Z}}, (w^j)_{\mathbf{Z}}).$$

Here $u(j,\cdot) = v^j$. We can go on to show A is closed, and has a bounded inverse $B: H \to H$. The abstract problem has the form

$$dp/dt + Ap = F(t,p), \quad p(0) = p_0,$$

where $p = (v,(v^j),(w^j))$. By obtaining some apriori estimates and applying a result in Friedman(1969), we have

Theorem 1 *Given the conditions of f stated above, if $(\varphi, \psi, h) \in X$, the problem (3.1), (3.2) has a unique solution for all $t > 0$.*

For steady state solutions, consider the case where

(3.4) $$f(u) = u(u-\alpha)(1-u), \quad 0 < \alpha < 1,$$

and consider

(3.5)
$$u_{xx} - gu = 0 \qquad x \in \mathbf{R}\backslash\mathbf{Z}$$

$$[[u_x]]_n + f(u) - \sigma u/\gamma = 0 \quad x = n \in \mathbf{Z}.$$

The general solution is given by

(3.6) $\quad u(x) = \{u_{n+1} \sinh(\sqrt{g}(x-n)) + u_n \sinh(\sqrt{g}(n+1-x))\}/\sinh\sqrt{g}$

$$\text{for } n \leq x \leq n+1, \, n \in \mathbf{Z},$$

where $u_n = u(n)$ satisfies

(3.7) $\quad\quad\quad\quad u_{n+1} - 2u_n + u_{n-1} + F(u_n) = 0,$

with

(3.8) $\quad F(u) = u\{2(1-\cosh\sqrt{g}) + (\sinh\sqrt{g}/\sqrt{g})[(u-\alpha)(1-u) - \alpha/\gamma]\}.$

The interesting case is when g is small, that is, the myelin resistivity is large, so that F has the qualitative form of f. For example, let $u_n = c > 0$ for all $n \in \mathbf{Z}$. Then letting $K \equiv \alpha + \sigma/\gamma + 2\sqrt{g}(\cosh\sqrt{g}-1)/\sinh\sqrt{g}$, c must satisfy (3.7), which can be written in the form

(3.9) $\quad\quad\quad\quad G(c) \equiv c^2 - (1+\alpha)c + K = 0.$

For two distinct (real) zeros of G, we must have

(3.10) $\quad\quad (1-\alpha)^2 > 4(\sigma/\gamma) + 8\sqrt{g}(\cosh\sqrt{g}-1)/\sinh\sqrt{g}.$

This holds if g is sufficiently small, and α, σ, γ are such that $(1-\alpha)^2 > 4\sigma/\gamma$. Given these conditions, there are two zeros $c = c_1, c_2$ of (3.9), $0 < c_1 < c_2 < 1$, and so we have at least two steady state solutions to (3.5), namely $u_i(x) = c_i R(x), i = 1, 2$, where

(3.11) $R(x) \equiv [\sinh\sqrt{g}(x-n) + \sinh\sqrt{g}(n+1-x)]/\sinh\sqrt{g}, \, n \leq x \leq n+1.$

Note that once $u(x)$ is determined, $w(n) = \sigma u(n)/\gamma$.

To consider the stability of steady state solutions to (3.1), suppose $(Q(x), (Q^j)_j, (Z^j)_j)$ is a steady state solution to (3.1), $Q^j = \lim_{x \to j} Q(x)$, and define

$$V(x,t) \equiv v(x,t) - Q(x) \quad (x \in \mathbf{R}), \, W(n,t) \equiv w(n,t) - Z^n \quad (n \in \mathbf{Z}),$$

where $(v, (v(n,\cdot))_n, (w(n,\cdot))_n)$ is a solution to (3.1). Define

$E_0(t) \quad \equiv \quad (1/2) \int_{\mathbf{R}\backslash\mathbf{Z}} (V^2 + (V_x)^2) dx,$

$E(t) \quad \equiv (1/2) \int_{\mathbf{R}\backslash\mathbf{Z}} (V^2 + a(V_x)^2) dx + (b/2) \sum_n \{(W(n,t))^2 + \sigma(V(n,t))^2\},$

where

$$a \equiv \min\{1, \gamma/\sigma, 2/B\}, \, b \equiv (ag+1)/\sigma.$$

We assume here that $f \in C^1(\mathbf{R})$ can be any function such that there is an interval $(\beta_1, \beta_2) \subset \mathbf{R}$, and constant $B > 0$ such that $-B \leq f'(y) < 0$ whenever $\beta_1 < y < \beta_2$. This is the B appearing in the definition of a. We also assumed the following conditions: v, w, Q satisfy

i) v is C^3 for $x \in \mathbf{R}\backslash \mathbf{Z}$, is C^0 for $x \in \mathbf{R}$, and is C^1 for $t > 0$, while $v_t \in C^0$ for $x \in \mathbf{R}$, and Q is C^3 for $x \in \mathbf{R}\backslash \mathbf{Z}$, is C^0 for $x \in \mathbf{R}$, w is C^1 for $t \in \mathbf{R}^+$ for each $n \in \mathbf{Z}$;

ii) there exists a $\delta > 0$ such that $-B \leq f'(y) \leq -a_1 < 0$ whenever $y \in (\beta_1+\delta, \beta_2 - \delta)$, and $v(x,0)$ satisfies $|v(x,0) - Q(x)| < \delta/2$;

iii) $\dot{E}(t), E'(t)$ are uniformly convergent on $\mathbf{R}$ and suppose

$$(E(0))^{1/2} < \sigma\sqrt{a}/2K,$$

where K is any constant which satisfies

$$\sup_{x \in \mathbf{R}} |v(x,t) - Q(x)| < K\sqrt{E_0(t)}.$$

Then,

Theorem 2 *If v, w, Q satisfy the above conditions, then $v(x,t) \to Q(x)$ uniformly on $\mathbf{R}$ as $t \to \infty$, and $w(n,t) \to Z^n$ as $t \to \infty$.*

We outline the proof here, but for details, consult Chen and Bell(1988). By (3.1)–(3.2) and definitions, V and W satisfy

$$V_t = V_{xx} - gV \qquad x \in \mathbf{R}\backslash \mathbf{Z}$$

$$\left.\begin{array}{l} V_t = [[V_x]]_n + F'(Q + \theta(u-Q)) - W \\ W_t = \sigma V - \gamma W \end{array}\right\} \quad x = n \in \mathbf{Z}$$

where $\theta = \theta(x,t) \in (0,1)$. By the definitions of a, b, B, δ, and F, we can find $p > 0$ such that

$$ax^2/2 + F'(\xi)xy + \sigma b(-F'(\xi))y^2 \geq p(x^2 + y^2)$$

$$ax^2/2 - axy + b\gamma y^2 \geq q(x^2 + y^2)$$

for all $(x,y) \in \mathbf{R}^2$, where $\xi \equiv Q + \theta(u-Q)$. With this one can show

$$E'(t) \leq -g\int_{\mathbf{R}\backslash \mathbf{Z}}(V^2 + aV_x^2)dx - \sum_{\mathbf{Z}}\{pV^2 + qW^2\}.$$

Choose a constant $r > 0$ such that $r \leq \min\{2p\sigma/(ag+1), q, 2g\}$. Then we have $E' \leq -rE$. With $R(x,t) = |V(x,t)|$ and $R(x,0) < \delta/2$, one then shows $R(x,t) < \delta/2$ for all $t \in \mathbf{R}^+$, all $x \in \mathbf{R}$. Also $F'(\xi(x,t)) \leq -a_1 < 0$ for all $x \in \mathbf{R}$, all $t \in \mathbf{R}^+$, so that $E'(t) \leq -rE(t)$ for all $t \in \mathbf{R}^+$, which implies $E(t) \to 0$ as $t \to \infty$. Since $\sup_x |u(x,t) - Q(x)| \leq K(E_0(t))^{1/2}$, then $u(x,t) \to Q(x)$ uniformly on $\mathbf{R}$, as $t \to \infty$. Also $w(x,t)$ converges since $[(ag+1)/2\sigma]|w(n,t) - Z^n|^2 \leq \{E(t)\}^{1/2}$.

As an example, consider the function $f(u) = u(u-\alpha)(1-u)$, where $0 < \alpha < 1$, and $0 \leq u \leq 1$. Now $f''(u) = 0$ when $u = (1+\alpha)/3$, and so we have $f'((1+\alpha)/3) = \max\{f'(x)|0 \leq x \leq 1\} = (1+\alpha^2 - \alpha)/3 > 0$. The two roots of $f'(u) = 0$ are

$$u_1 = \alpha/(1 + \alpha + (1+\alpha^2 - \alpha)^{1/2})$$

$$u_2 = (1 + \alpha + (1+\alpha^2 - \alpha)^{1/2})/3.$$

If we choose $F(u) = f(u)$ in Theorem 2, then $F'(u) < 0$ whenever $u \in (0, u_1)$ or $u \in (u_2, 1)$. Hence we have

Corollary 3 *If we choose $F(u) = u(u-\alpha)(1-u)$, $0 < \alpha < 1$, and for the interval $(a, b) = (0, u_1)$ or $(a, b) = (u_2, 1)$, then we have $u(x, t) \to Q(x)$ uniformly on $\mathbf{R}$ as $t \to \infty$, and $w(n, t) \to (\sigma/\gamma)Q(n)$ on $\mathbf{Z}$ as $t \to \infty$.*

For example, if we choose $\alpha = 2/3$, $\sigma/\gamma = 1/40$, and g small enough, there exists a $u_{2g}(x)$ which lies entirely in $(u_2, 1)$. In this case $u_2 = (5 + \sqrt{7})/9 \cong 0.8495279$. From (3.9), we have the larger root

$$c_2(g) = (5/3 + (1/90 - 8G(\cosh\sqrt{g} - 1))^{1/2})/2.$$

When $g = 0$, $c_2(0) > 0.88603 > u_2$. Hence, we can choose g small enough such that $1 > c_2(g) > u_2$. That means there is a $g^* > 0$ such that the above inequality holds whenever $0 \leq g < g^*$. Then

$$Q(x) = c_2(g)(\sinh\sqrt{g}(x-j) + \sinh\sqrt{g}(j+1-x))/\sinh\sqrt{g}, \quad j \leq x \leq j+1$$

is a solution to (3.5). Since

$$\min\{Q(x)|\, j \leq x \leq j+1\} = Q(j+1/2) = c_2(g)/\cosh\sqrt{g}/2,$$

there exists a g_* which satisfies $g^* \geq g_* > 0$ such that

$$\min\{Q(x)|\, j \leq x \leq j+1\} > u_2 \text{ whenever } 0 \leq g \leq g_*.$$

Since $\max\{Q(x)|\, j \leq x \leq j+1\} = Q(j) = Q(j+1) = c_2(g) < 1$, we conclude that $Q(x) \in [u_2, 1]$ for all $x \in \mathbf{R}$.

Without attacking the problem numerically it is rather difficult to refine the threshold conditions on the problem (3.1)–(3.2). In earlier studies (Bell and Cosner 1983) we considered the initial value problem without the recovery variable. That is, the problem addressed in (Bell and Cosner 1983) was

(3.12) $$v_t = v_{xx} - gv \quad x \in \mathbf{R}\backslash\mathbf{Z}$$

(3.13) $$v_t = [[v_x]]_n + f(v) \quad x = n \in \mathbf{Z}$$

where f has the form given in Figure 2. In this case problem (3.12)–(3.13) has a comparison principle. Again considering steady state solutions to (3.12)–(3.13), say $q = q(x)$ with $q(n) = q_n$, then $\{q_n\}_{n \in \mathbf{Z}}$ satisfies

(3.14) $$q_{n+1} - (2\cosh\sqrt{g})q_n + q_{n-1} + (\sinh\sqrt{g}/\sqrt{g})f(q_n)$$
$$= q_{n+1} - 2q_n + q_{n-1} + F_g(q_n) = 0,$$

where
$$F_g(q) \equiv 2(1 - \cosh\sqrt{g})q + (\sinh\sqrt{g}/\sqrt{g})f(q).$$

If one looks for special steady states where $q_n = $ constant $= c$ for all $n \in \mathbf{Z}$, then there exists a $g^* > 0$ such that for $0 < g < g^*$, there are two c's, that is, two constant solutions to (3.14), $c = c_a, c_b$, where $0 < c_a < c_b$. If we call the steady state $Q(x)$, which satisfies $Q(j) = c_b$, the fully activated state of (3.14), then

(3.15) $\qquad Q(x) = [\sinh\sqrt{g}(x - j) + \sinh\sqrt{g}(j + 1 - x)](c_b/\sinh\sqrt{g}),$

for $j < x < j + 1, \; j \in \mathbf{Z}$.

The main technical lemma we have to use on initial value problems of the form (3.12)–(3.13) is the following statement, which is a "myelinated axon model" analogue to proposition (2.2) given by Aronson and Weinberger (1975).

Lemma 4 *Suppose for $x_1 < x_2$ that $q(x) \in [0,1]$ is a steady state solution to (3.12)–(3.13) for $x \in (x_1, x_2)$, with $q(x_i) = 0$ if x_i is finite, $i = 1, 2$. Let $v(x,t)$ be continuous in x and t, C^1 in t for all x, and C^2 in x for $x \notin \mathbf{Z}$, and suppose v satisfies (3.12)–(3.13) with*

$$v(x,0) = \begin{cases} q(x) & x \in (x_1, x_2) \\ 0 & x \notin (x_1, x_2). \end{cases}$$

Then, for each $x \in \mathbf{R}$, $v(x,t)$ is a nondecreasing function of t, with $\lim_{t \to \infty} v(x,t) = \tau(x)$, converging uniformly in each bounded interval, where $\tau(x)$ is the smallest nonnegative steady state solution to (3.12)–(3.13) for $x \in \mathbf{R}$ which satisfies $\tau(x) \geq q(x)$ for $x \in (x_1, x_2)$.

For the proof, see Bell and Cosner(1983). It relies on a comparison principle for the system so there is no result of similar strength that has been obtained for the model with recovery. An application of the lemma to be stated now was the strongest stability statement we made in (Bell and Cosner 1983), and was phrased in terms of $F_g(v)$ defined above, to which we refer the reader for details. Let $M_g \equiv \sup\{F_g(v)|\, 0 \leq v \leq c_b\}$ and $m_g \equiv -\inf\{F_g(v)|\, 0 \leq v \leq c_b\}$.

Theorem 5 *Suppose there exist constants α, β, with $c_a < \alpha < \beta < c_b$ such that $2v - F_g(v) = 0$ for $v = \alpha, \beta$, and $2v - F_g(v) < 0$ for $\alpha < v < \beta$. Let $p(x) = [\alpha \sinh\sqrt{g}(1 - x\,\mathrm{sgn}(x))]/\sinh\sqrt{g}$. Assume at least one of the following conditions holds:*

$$2\alpha + m_g \leq \beta;$$

$$\alpha + c_b + M_g \leq 2\beta.$$

If $v(x,t)$ is a solution to (3.12)–(3.13) with $v(x,0) \geq 0$ for all x, and for some ζ, $v(x + \zeta, 0) \geq p(x)$ for $|x| \leq 1$, then $\lim_{t \to \infty} v(x,t) = Q(x)$, for all $x \in \mathbf{R}$, where Q is defined by (3.15).

Finally, through comparison principles, bounds on the speed of propagation of travelling wave "fronts" (since there is no recovery processes in the problem) were obtained in (Bell and Cosner 1983).

Some work has been initiated to obtain (small amplitude) travelling wave-like solutions to the model with recovery. This work is still preliminary, and our general analytical knowledge about action potentials, that is, travelling wave solutions, is essentially nonexistent.

For vanishingly small internodes the above models do not converge in form to the unmyelinated case, so Grindrod and Sleeman (1985) proposed another model. By their approach the analogue model to system (3.1) is

(3.16) $$\begin{cases} cv_t + gv = v_{xx} & x \in (\theta, 1) \bmod 1 \\ v_t - f(v) + w = \beta v_{xx} \\ w_t = \sigma v - \gamma w \end{cases} \quad x \in (0, \theta) \bmod 1$$

where the nodes constitute the intervals $(n, n + \theta), n \in \mathbf{Z}$, of width $\theta \in (0, 1)$. Interface conditions are needed, and they become

(3.17) $$[[v]]_y = [[v_x]]_y = 0 \quad y \in 0, \theta \bmod 1.$$

That is, there is continuity of potential and current across the interfaces between myelinated and unmyelinated regions. Grindrod and Sleeman (1985) considered the model without a recovery variable, and with use of a comparison principle were able to obtain long time behaviour results.

With the presense of a recovery variable w, Lyapunov methods are again appropriate to examine stability conditions on steady state solutions. For convenience, take the case $c = \beta = 1$, let f be C^1 on $\mathbf{R}$ such that there exists an interval (a_0, b_0) and a constant $B_0 > 0$ such that $-B_0 \leq f'(y) < 0$ for $y \in (a_0, b_0)$, and f' is uniformly bounded on $\mathbf{R}$. Suppose $Q(x)$ is a solution to the steady state problem

(3.18) $$Q_{xx} + f(Q) - \sigma Q/\gamma = 0 \quad x \in (0, \theta) \bmod 1$$

$$Q_{xx} - gQ = 0 \quad x \in (\theta, 1) \bmod 1$$

(3.19) $$[[Q]]_y = [[Q_x]]_y = 0 \quad y = 0, \theta \bmod 1$$

with $Q(x) \in (a_0 + \delta, b_0 - \delta)$ for all $x \in \mathbf{R}$ for some $0 < \delta < (b_0 - a_0)/2$. Define

(3.20) $$u = v - Q, \quad z = w - \sigma Q/\gamma,$$

and $F(v) \equiv \int \{f(v) - f(Q)\} dy$ (the integral taken over $(Q, Q + v)$)

$$E_0(t) \equiv (1/2) \int_{\mathbf{R}} (u^2 + u_x^2) dx$$

(3.21) $$E(t) \equiv (1/2) \sum_n \{ \int_n^{n+\theta} (u^2 + 2aF(u) + au_x^2 + bz^2 + ag(u - z/g)^2 + cz_x^2) dx + \int_{n+\theta}^{n+1} (u^2 + au_x^2) dx \}.$$

Here a, b, c are positive constants to be determined.

Theorem 6 *Assume a solution (v, w) to (3.16)–(3.17) the "energy" function $E(t)$ defined in (3.21), and positive constants σ, g, γ, a, b,c satisfy*

a) $v \in C^1(\mathbf{R} \times \mathbf{R}^+)$;
b) E, E' *are uniformly convergent on* $\mathbf{R}^+$ *and suppose* $(E(0))^{1/2} < \delta\sqrt{a}/2\sqrt{2}$;
c) $\sigma < g < \gamma/B_1$, *where* $B_1 = \max\{2, \sup_{y \in \mathbf{R}} |f'(y)|\} < \infty$; a *and* c *in the definition of E satisfy*

$$0 < a < \min\{1/B_1, 1/2(\gamma + \sigma/g - g)\}, \quad 0 < c < 2/\sigma,$$

$$b\sigma = 1 + ag - a\gamma - a\sigma/g.$$

Then $v(x,t) \to Q(x)$ uniformly on $(0, \theta)$ mod 1 as $t \to \infty$ for each $n \in \mathbf{Z}$.

A proof of this appears in (Chen, unpublished thesis, 1988), but the argument is very similar to that employed in the proof of Theorem 2. One has to show $E'(t) < -rE(t)$ for all $t \in \mathbf{R}^+$, for some $r > 0$. From there it is easy to conclude $E(t) \to 0$ as $t \to \infty$. Since $\sup_x |u(x,t)| \leq (2E_0(t))^{1/2}$, then $u(x,t) \to 0$ uniformly in $\mathbf{R}$ as $t \to +\infty$.

There are a variety of such conditions one can impose to obtain stability results, depending on the nature of the nonlinearity. In another direction, let f be of the type considered in the first model, i.e., f be C^1, and there is an $\alpha \in (0,1)$ such that $f(v) < 0$ for $v \in (0, \alpha)$, $f(v) > 0$ for $v \in (\alpha, 1)$, $f(0) = f(\alpha) = f(1) = 0$, $f'(v) < 0$ for $v \in (-\infty, 0) \cup (1, \infty)$, and f' remains bounded on $\mathbf{R}$. One might consider steady state solutions, that is (nontrivial) solutions $(Q(x), Z(x) = \sigma Q(x)/\gamma)$ to (3.18)–(3.19), which satisfy

(3.22) $\qquad\qquad Q(x+1) = Q(x)$ for all $x \in \mathbf{R}$.

That is, we look for 1-periodic solutions to (3.18)–(3.19).

Theorem 7 *(A) If the constants g, σ, γ and function f are fixed, then there exists a $\theta^0 > 0$ such that if $0 < \theta < \theta^0$, then there is no solution $Q(x)$ to (3.18)–(3.19) satisfying (3.22).*

(B) For any $0 < \theta_0 < 1$, there exists a 1-periodic solution to (3.18)–(3.19) with node length θ satisfying $\theta_0 < \theta < 1$.

Statement (A) constitutes Theorem 1.3 in (Grindrod and Sleeman 1985). For both cases it is a matter of considering the behaviour of two "time maps" since there are two dynamical systems to analyze. In fact, with only a slight refinement of argument of (B), one can find a sequence $\{\theta_k\}$ which satisfies $\theta_k \in (0,1)$, $\lim_{k \to \infty} \theta_k = 1$, such that for each k, there is a 1-periodic solution to (3.18)–(3.19) with node length θ_k. For details of the proof of Theorem 7, see Grindrod ans Sleeman (1985), and Chen (1988). In fact, Chen (1988) proves

Theorem 8 *There exists a $0 < \theta^* < 1$ such that for any $\theta \in (\theta^*, 1)$, there exists a 1-periodic solution to (3.18)–(3.19) with nodal length θ.*

The direction the work on diffusion models could take, analytically, would be to explore threshold conditions with the type of current stimuli (boundary data) discussed at the beginning of the section, as well as explore conditions that support travelling wave-like solutions.

In the next section we take a different tack. We turn to a spatially discrete model approach to a myelinated axon and discuss one aspect of solution behaviour for a new class of models.

4 Spatially discrete models

The history of quantitative models for a single myelinated axon predates the work of FitzHugh, but earlier work generally considered differential-difference models of these axons. See, for example, Landahl and Podolsky (1949), Scott (1967), Picard (1969), Bean (1974), McNeal (1976), Bell (1981), Bell and Cosner (1984). The general assumption behind such models, besides the previously mentioned hypotheses of negligible cross-sectional variation in potential, and nodes being point sources of excitation, is that the myelin provides such high resistance and low capacitance compared to nodal membrane that it completely insulates the membrane. Thus a circuit model like that given in Figure 4 is replaced by one where there is only a single circuit element representing any one node, and no current pathways between nodes. If $v_k(t)$ represents the transmembrane potential at node k at time t, these models take the form

$$(4.1) \qquad RC dv_k/dt + I_{ion}(v_k) = v_{k+1} - 2v_k + v_{k-1},$$

where a variety of current-voltage relations, $I_{ion}(\cdot)$, have been employed to examine some behaviour of solutions to (4.1). For example, using a linear current-voltage relation, Landahl and Polodsy (1949) showed there was a critical internodal length and diameter for their model that maximized conduction velocity, while McNeal (1976) computed strength-duration curves using a more physiologically realistic I_{ion}. Some aspects of conduction behaviour were considered by Scott (1967), Picard (1969), and Bell (1981). Bell (1981) and Bell and Cosner (1984) also considered threshold behaviour in a model like (4.1) with I_{ion} being a cubic-like function as shown in Figure 2 and with no recovery terms. The main tool in these studies was again a comparison principle for (4.1). This class of problems has served as models in other contexts. For example, Keener (1987, 1988) has used this type of system to model propagation failure in myocardial tissue.

Since the electrophysiological properties of the internode are very important in propagation of action potentials, the class of problems (4.1) is too simple to serve as a reasonable model for myelinated fibre, where the lipoprotein wrapping is not tight, and where there is evidence of ionic channel distribution in the internodes. The next simplest model, spatially, is to consider a circuit model like that in Figure 5. That is, myelin membrane 'elements' are alternated between nodal membrane elements of two types, respectively represented by the notation 'N' and 'M'. We refer to this circuit and associated mathematical model as the lumped myelinated segment (LMS) model. If $v_k = v_k^i - v_k^0$ and $u_k = u_k^i - u_k^0$ represent the membrane potential at the kth node and k^{th} myelin segment, respectively, then the equations derived from the circuit model, Figure 5, have the form

$$(4.2) \quad \begin{cases} C_M du_k/dt + \tilde{I}(u_k) = (2/R)(v_{k-1} - 2u_k + v_k) \\ C_N dv_k/dt + I(v_k) = (2/R)(u_k - 2v_k + u_{k+1}). \end{cases}$$

This model was introduced in (Bell 1984), but only recently has the author returned to look at the model (4.2) further. Below we will outline some of the work done on the model, where FitzHugh-Nahumo dynamics is imposed at the nodes.

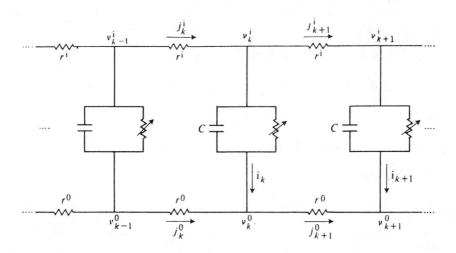

Figure 5. Circuit model for equation (4.1), with $v_k = v_k^i - v_k^0$.

As we considered for the diffusion model above, since myelin resistivity is large compared with nodal membrane, the membrane current across the myelin elements is modelled electrotonically. That is, let $\tilde{I}(u) = u/r$, where the constant r represents the myelin resistivity. After substitution of FitzHugh-Nagumo dynamics at the nodes, as was done for (2.4), and some rescaling done , (4.2) takes the form

(4.3)
$$\begin{cases} Cdu_k/dt + gu_k = d(v_{k-1} - 2u_k + v_k) \\ dv_k/dt + f(v_k) + w_k = d(u_k - 2v_k + u_{k+1}) \\ dw_k/dt = \sigma v_k - \gamma w_k \end{cases} \quad (k \in \mathbf{Z},\ t > 0),$$

where $g = 1/rC_N, C = C_M/C_N$, and $d = 2/RC_N$, as well as σ and γ, are fixed positive parameters.

The following two results deal with the stability of the trivial steady state, i.e., $(u_k, v_k, w_k) = (0, 0, 0)$ for all $k \in \mathbf{Z}$. We assume the positive constants C, g, d, σ and γ are fixed and also assume f is given by

(4.4) $\quad f(v) = f_0 v(v - a)(v - b), \quad f_0, a, b$ are positive, and $b/2 < a < b$.

We are taking too restrictive a function f, but it is used to illustrate the proof. We will make a remark later concerning what conditions we need on a general f for Theorems 9 and 10 to hold. First we show that $(u_k, v_k, w_k) = (0, 0, 0)$ is locally stable. That is, if the initial data is "small", and the "initial energy" imposed on the system is "small", where both concepts are specified more precisely below, then $\lim_{t \to \infty}(u_k(t), v_k(t), w_k(t)) = (0, 0, 0)$. Let

(4.5)
$$\begin{aligned} E_1(t) &\equiv \sum_j \{(kd/2)(u_j - v_j)^2 + (kd/2)(v_j - u_{j+1})^2 + kF(v_j) \\ &\quad + (m/2)(v_j)^2 + (k\mu/2)(u_j)^2 + (k/2)(v_j + w_j)^2 \\ &\quad + K/2[(v_j)^2 + (w_j)^2]\}, \end{aligned}$$

and

(4.6) $$E_2(t) \equiv \sum_j \{(u_j)^2 + (v_j)^2 + (w_j)^2\}.$$

Also let

(4.7)
$$\begin{aligned} r &= 2 + kf_0[2(a+b)^2/9 - ab + (1 + \gamma - \sigma)/f_0] \\ s &= (k + \sigma^{-1})/abf_0 + k(\gamma\sigma^{-1} - 1) \\ S &= (\sigma g)^{-1} - k(1 - \gamma\sigma^{-1}) \\ \alpha &= (a+b)/2 - \{(b-a)/2 + (1 + \sigma k)/f_0(K + k + m)\}^{1/2}. \end{aligned}$$

We assume the following relations hold:

(4.8)
$$\begin{cases} 0 < k < \sigma/\gamma^2 \\ m \equiv K(\sigma - 1) + k(\sigma - 1 - \gamma) \\ \mu \equiv g + (K + k + m)C/k \\ K > \max\{2, 2/C, \gamma/\sigma, \gamma^{-1} - k, s, S\}. \end{cases}$$

With f given by (4.4), we also define

(4.9) $$F(v) = \int_0^v f(z)\,dz = f_0 v^2\{v^2/4 - (a+b)v/3 + ab/2\}.$$

Theorem 9 *Suppose $\{(u_k, v_k, w_k)\}_{\mathbf{Z}}$ is a solution to (4.3) such that for some $t_0 \geq 0$, $\sup_j |v_j(t_0)| < \alpha$ and $(E_1(t_0))^{1/2} < \alpha$. Then $\lim_{t \to \infty}(u_k(t), v_k(t), w_k(t)) = (0,0,0)$ for all $z \in \mathbf{Z}$.*

Proof: Upon substituting E_2 from E_1, we have

$$\sum_j\{(kd/2)(u_j - v_j)^2 + (kd/2)(v_j - u_{j+1})^2 + kF(v_j) + (m/2)(v_j)^2$$

$$+ (k/2)(v_j + w_j)^2 + (k\mu/2)(u_j)^2$$

$$+ (K/2)[(v_j)^2 + (w_j)^2] - (u_j)^2 - (v_j)^2 - (w_j)^2\}$$

$$= \sum_j\{(kd/2)(u_j - v_j)^2 + (kd/2)(v_j - u_{j+1})^2$$

$$+ kf_0(v_j)^2[(v_j)^2/4 - (a+b)v_j/3 + (ab/2 + (m+K-2)/2kf_0)]$$

$$+ (k\mu - 2)(u_j)^2/2 + (K - 2)(w_j)^2/2 + (k/2)(v_j + w_j)^2\}$$

which is positive by (4.8) if the $(v_j)^2$ term is nonnegative, since $K > 2$, $k > 0$, and $CK > 2$, so that $k\mu > 2$. The $(v_j)^2$ term is positive if the discriminant associated with the quadratic expression is negative, and this is so if $m + K > 2 + kf_0[2(a+b)^2/9 - ab]$, which holds since $\sigma K > r$, r defined in (4.7). Hence $E_1(t) \geq E_2(t)$. Now taking the derivative of $E_1(t)$, we obtain

$$\dot{E}_1(t) = dE_1(t)/dt$$

$$= \sum_j\{kd(u_j - v_j)(\dot{u}_j - \dot{v}_j) + kd(v_j - u_{j+1})(\dot{v}_j - \dot{u}_{j+1}) + kf(v_j)\dot{v}_j +$$

$$mv_j\dot{v}_j + k(v_j + w_j)(\dot{v}_j + \dot{w}_j) + k\mu u_j\dot{u}_j + K[v_j\dot{v}_j + w_j\dot{w}_j]\}.$$

Since, from (4.3),

$$\sum_j\{(u_j - v_j)(\dot{u}_j - \dot{v}_j) + (v_j - u_{j+1})(\dot{v}_j - \dot{u}_{j+1})\}$$

$$\begin{aligned}
&= \sum_j \{-\dot{u}_j(v_j - u_j) - \dot{v}_j(u_j - 2v_j + u_{j+1}) - \dot{u}_j(v_{j-1} - u_j)\} \\
&= d^{-1}\sum_j \{-\dot{v}_j(\dot{v}_j + f(v_j) + w_j) - \dot{u}_j(C\dot{u}_j + gu_j)\} \\
&= d^{-1}\sum_j \{-(\dot{v}_j)^2 - C(\dot{u}_j)^2 - \dot{v}_j f(v_j) - \dot{v}_j w_j - gu_j\dot{u}_j\},
\end{aligned}$$

then

$$\begin{aligned}
\dot{E}_1 &= \sum_j \{-k(\dot{v}_j)^2 - kC(\dot{u}_j)^2 + \sigma k(v_j)^2 - \gamma(k+K)(w_j)^2 \\
&\quad + [\sigma(k+K) - k\gamma]v_j w_j + d(m+k+K)v_j(u_j - 2v_j + u_{j+1}) \\
&\quad + (kd/C)(\mu - g)u_j(v_{j-1} - 2u_j + v_j) - (kg/C)(\mu - g)(u_j)^2 \\
&\quad - (m+k+K)v_j(f(v_j) + w_j)\}.
\end{aligned}$$

Noting that

$$\sum_j \{v_j(u_j - 2v_j + u_{j+1}) + u_j(v_{j-1} - 2u_j + v_j)\}$$
$$= -\sum_j \{(v_j - u_{j+1})^2 + (u_j - v_j)^2\},$$

and using the definition of μ in (4.8), we have

$$\begin{aligned}
\dot{E}_1 &= -E_2 + \sum_j \{-k(\dot{v}_j)^2 - kC(\dot{u}_j)^2 - d(m+k+K)(u_j - v_j)^2 \\
&\quad - d(m+k+K)(v_j - u_{j+1})^2 - [\gamma(k+K) - 1](w_j)^2 \\
&\quad - [g(m+k+K) - 1](u_j)^2 \\
&\quad - (m+k+K)[v_j f(v_j) - (\sigma k + 1)(v_j)^2/(m+k+K)]\}.
\end{aligned}$$

It follows that the first two square-bracketed terms are positive since $K > \gamma^{-1} - k$ and $K > S$, since $k < \sigma/\gamma^2$ implies $m + k + K > 0$. The third square-bracketed term has the form $(v_j)^2 P(v_j)$, where P has real roots, the smallest being α. Since $\alpha > 0$, if $K > s$, then

(4.10) $$\dot{E}_1(t) < -E_2(t) \le 0,$$

provided that each $v_j \le \alpha$. By hypothesis, $\dot{E}_1(t_0) < 0$, and so we claim $\dot{E}_1(t) \le 0$ for all $t \ge t_0$. Suppose there is a $t^* > t_0$ such that for some i, $v_i(t^*) > \alpha$, and let $T = \inf\{t > t_0 \mid v_i(t) > \alpha \text{ for some } i\}$. Then for $t \le T$, $v_j(t) \le \alpha$ for all j, so that $\dot{E}_1(t) \le 0$ for all $t \le T$. Therefore $E_1(T) \le E_1(t_0)$. But, since $\sup_j |v_j(t)| \le [E_2(t)]^{1/2} \le [E_1(T)]^{1/2} < \alpha$, This is a contradiction. Hence $E_1(t)$ is a bounded, nonincreasing, positive function for $t > t_0$, so $\lim_{t\to\infty} E_1(t) = \varepsilon$ exists.

If $\varepsilon = 0$, then the theorem's conclusion follows because $\lim_{t \to \infty} E_2(t) = 0$. Since E_2 is bounded, the v'_js, w'_js and u'_js are also bounded, so that there is an $M > 0$ such that $E_1(t) \leq M E_2(t)$. This, along with (4.10), implies $\dot{E}_1(t) \leq -E_1(t)/M$ for $t > t_0$. Hence $E_1(t)$ decays to zero.

Remark) For K sufficiently large, $\alpha - a = O(K^{-1})$, where a is as in (4.4), so that the conclusion of Theorem 9 still holds if we replace α in the hypotheses by $\sup_j |v_j(t_0)| < a$ and $[E_1(t)]^{1/2} < a$.

Remark) As previously mentioned, using the nonlinearity is too restrictive. We can actually replace (4.4) by

(4.11) f is a smooth function such that there are constants $0 < a < b$ such that $f(0) = f(a) = f(b) = 0$, with $f(v) > 0$ for $0 < v < a$, and $f(v) < 0$ for $a < v < b$, $f'(0) > 0$, and $\int_0^b f(v)\, dv < 0$.

Then for k, K^{-1} sufficiently small, $kF(v) + (m+k+K)(v)^2/2 \geq 0$, which is needed for $E_1(t) \geq E_2(t)$, and for $\dot{E}_1(t) \leq -E_2(t)$, we would want $vf(v) - (\sigma k + 1)v^2/(m+k+K) > 0$ for $0 < v < \alpha$, for some α. This would always hold for (4.11), with $f'(0) > 0$. That is, using (4.11) we would let α br the smallest positive zero of

$$f(v) - (\sigma k + 1)v/(m + k + K) = 0,$$

then the theorem's conclusion follows for K^{-1}, k sufficiently small.

In the following we show that $(u_k, v_k, w_k) \equiv (0, 0, 0)$ is not a globally stable solution to (4.3). Let

$$E_3(t) = \sum_j \{d(u_j - v_j)^2/2 + d(v_j - u_{j+1})^2/2 + F(v_j) + A(v_j)^2/2 \\ - Bv_j w_j + D(w_j)^2/2 + g(u_j)^2/2\}.$$

Assume the following relations hold:

(4.12) $$\gamma^2 \geq \sigma$$

(4.13)
$$A = 2(\gamma + \sqrt{\gamma^2 - \sigma}),$$
$$B = A^2/4\sigma = A\gamma/\sigma - 1,$$
$$D = (B+1)^2/4\gamma = A^2\gamma/4\sigma^2.$$

Theorem 10 *If for some $t_0 \geq 0$, $E_3(t_0) < 0$, then for each $t \geq t_0$, there is an $i \in \mathbf{Z}$ such that $F(v_i(t)) < 0$, which implies $v_i(t) > a$.*

Proof: Differentiating E_3 and using (4.3), we have

$$\dot{E}_3 = \sum_j \{-(\dot{v}_j)^2 - C(\dot{u}_j)^2 - \sigma B(v_j)^2 - \gamma D(w_j)^2 \\ + (B\gamma + D\sigma)v_j w_j + Av_j \dot{v}_j - (1+B)\dot{v}_j w_j\}.$$

Since

$$-(\dot{v}_j - Av_j/2 + (1+B)w_j/2)^2 = -(\dot{v}_j)^2 - A^2(v_j)^2/4 - (1+B)^2(w_j)^2/4 \\ + Av_j\dot{v}_j + A(1+B)v_j w_j/2 - (1+B)\dot{v}_j w_j,$$

we can write $\dot{E}_3$ as

$$\dot{E}_3 = \sum_j \{-(\dot{v}_j - Av_j/2 + (1+B)w_j/2)^2 + [(1+B)^2/4 - D\gamma](w_j)^2 \\ + [A^2/4 - B\sigma](v_j)^2 + [D\sigma + B\gamma - A(1+B)/2]w_j v_j - C(\dot{u}_j)^2\}.$$

By using the definition of A, B, and D in (4.13), the square-bracketed terms are zero, so that

$$\dot{E}_3 = -\sum_j \{(\dot{v}_j - Av_j/2 + (1+B)w_j/2)^2 + C(\dot{u}_j)^2\} \leq 0.$$

Thus given the theorem's hypothesis, $E_3(t) < 0$ for all $t \geq t_0$. If we let

$$\mathcal{X}(u, v, w) \equiv gu^2/2 + Av^2/2 - Bvw + Dw^2/2,$$

then by (4.12)–(4.13),

$$\mathcal{X}(u, v, w) = gu^2/2 + \sigma v^2/2\gamma + D(w - \sigma v/\gamma)^2/2 > 0,$$

so if $E_3(t) < 0$, then

(4.14)
$$0 > \sum_j \{d(u_j - v_j)^2/2 + d(v_j - u_{j+1})^2/2 + F(v_j) + \mathcal{X}(u_j, v_j, w_j)\} \\ \geq \sum_j F(v_j(t)).$$

Let $k > a$ be such that $F(k) = 0$. Then, by (4.14), $F(v_i(t)) < 0$ for some i, which, because of (4.9), means $v_i(t) > k > a$.

Remark) Since the proof theorem 10 does not use qualitative properties of f, or F, beyond the fact that F is initially positive and becomes negative somewhere, this theorem holds if (4.4) is replaced by (4.11).

An interesting question to consider for the class of models (4.3) is when such a model would possess travelling wave solutions. As was discussed in Bell (1984) for a simpler case, for such a solution we assume there exists a $\tau > 0$, called the internodal delay, such that

(4.15)
$$u_k(t) = u_{k+1}(t+\tau), \ v_k(t) = v_{k+1}(t+\tau), \ w_k(t) = w_{k+1}(t+\tau),$$
for all $k \in \mathbf{Z}$, and all $t \in \mathbf{R}$.

If (4.15) is used in (4.3), we may consider the potential at any one node, and any adjacent lumped myelinated segment, say the nth, and let $(\psi(t), \varphi(t), \omega(t)) = (u_n(t), v_n(t), w_n(t))$. Then (4.3) becomes

(4.16)
$$\begin{cases} Cd\psi(t)/dt &= d(\varphi(t+\tau) - 2\psi(t) + \varphi(t)) - g\psi(t) \\ d\varphi(t)/dt &= d(\psi(t) - 2\varphi(t) + \psi(t-\tau)) - f(\varphi(t)) - \omega(t) \\ d\omega(t)/dt &= \sigma\varphi(t) - \gamma\omega(t) \quad t \in \mathbf{R}. \end{cases}$$

A travelling wave for (4.3) would be a nontrivial solution to (4.16) satisfying $(\psi(t), \varphi(t), \omega(t)) \to (0,0,0)$ as $|t| \to \infty$. The existence of such a solution remains open at present.

References

[1] Aronson, D. G. and Weinberger, H. F.(1975). Nonlinear diffusion in population genetics, combustion, and nerve propagation. In *Proceedings of the Tulane Program in Partial Differential Equations and Related Topics*, (ed. J. A. Goldstein *Lecture Notes in Mathematics*, No. 446, Springer-Verlag, Berlin.

[2] Bean, C. P.(1974). A theory of microstructure of myelinated fibres. *Journal of Physiology*, **243**, 514-22.

[3] Bell, J.(1981). Some threshold results for models of myelinated nerves. *Mathematical Biosciences*, **54**, 181-90.

[4] Bell, J.(1984). Behavior of some models of myelinated axons, *IMA Journal of Mathematics Applied in Medicine and Biology*, **1**, 149-67.

[5] Bell, J. and Cosner, C. G.(1983). Threshold conditions for a diffusive model of a myelinated axon. *Journal of Mathematical Biology*, **18**, 39-52.

[6] Bell, J. and Cosner, C. G.(1984). Threshold behavior and propagation for nonlinear differential-difference systems motivated by modeling myelinated axons. *Quarterly Journal of Applied Mathematics*, **42**, 1-14.

[7] Bell, J. and Cosner, C. G.(1986). Threshold conditions for two diffusion models suggested by nerve impulse conduction. *SIAM Journal of Applied Mathematics*, **46**,844-55.

[8] Chen, P.(1988). Global solutions, long time behavior, and steady state solutions to myelinated nerve axon models with FitzHugh-Nagumo dynamics. unpublished Ph. D. Thesis, State University of New York at Buffalo.

[9] Chen, P. and Bell, J.(1988). Global solutions and long time behavior to a FitzHugh-Nagumo myelinated axon model. *SIAM Journal of Mathematical Analysis*, **20**, 567-81.

[10] FitzHugh, R.(1962). Computation of impulse initiation and saltatory conduction in a myelinated nerve fiber. *Biophysical Journal*, **2**, 11-21.

[11] Friedman, A.(1969). *Partial Differential Equations*. Holt, Rinehart, and Winston, New York.

[12] Goldman, L. and Albus, J. S.(1968). Computation of impulse conduction in myelinated fibres: theoretical basis of the velocity-diameter relation. *Biophysical Journal*, **8**, 596-607.

[13] Grindrod, P. and Sleeman, B.(1985). A model of a myelinated nerve axon: threshold behaviour and propagation. *Journal of Mathematical Biology*, **23**, 119-35.

[14] Hodgkin, A. L. and Huxley, A. F.(1952). A quantitative description of membrane current and its application to conduction and excitation in nerve. *Journal of Physiology*, **117**, 500-44.

[15] Keener, J.(1987). Propagation and its failure in coupled systems of discrete excitable cells. *SIAM Journal of Applied Mathematics*, **47**, 556-72.

[16] Keener, J.(1988). A mathematical model for the vulnerable phase in myocardium. *Mathematical Biosciences*, **90**, 3-18.

[17] Landahl, H. D. and Podosky, R. J.(1949). Space velocity of conduction in nerve fibres with saltatory transmission. *Bulletin of Mathematical Biophysics*, **11**, 19-27.

[18] McNeal, D. R.(1976). Analysis of a model for excitation of myelinated nerve. *IEEE Biomedical Engineering*, **23**, 329-37.

[19] Picard, W. F.(1969). Estimating the velocity of propagation along myelinated and unmyelinated fibres. *Mathematical Biosciences*, **5**, 305-19.

[20] Scott, A. C.(1967). More on myelinated nerve model analysis. *Mathematical Biophysics*, **29**, 363-71.

[21] Tuckwell, H. C.(1988). *Introduction to Theoretical Neurobiology*, Vol 2, Cambridge University Press.

[22] Waxman, S. G.(1972). Regional differentiation of the axon: a review with special reference to the concept of the multiplex neuron. *Brain Research.* **47**, 269-88.

Eigenvalue problems with indefinite weights and reaction-diffusion models in population dynamics

Chris Cosner
University of Miami

1 Introduction and motivation

The object of this article is to illustrate how the theory of linear elliptic eigenvalue problems with indefinite weight functions can be used to analyze reaction-diffusion models in mathematical ecology and population genetics. It is not intended to be a comprehensive survey, but rather an introduction to some of the principal ideas and methods. Reaction-diffusion equations arise in ecological modelling when nonlinear dynamics describing the growth or decline of a population are combined with a diffusion process describing the spatial dispersal of that population. Solutions to the reaction-diffusion models typically represent population densities, or in related problems of population genetics, the distribution of certain alleles within a population. In many cases, the behaviour of solutions of such reaction-diffusion equations is determined by the nature of the equilibrium states. Those in turn can often be described via such methods as bifurcation theory and linearized stability analysis, which immediately lead to problems in linear elliptic spectral theory. Both the modelling and the analysis can introduce considerations which require the study of elliptic eigenvalue problems with indefinite weight functions, and such problems will be the main subject of discussion.

The two main components of the models we consider are the "reaction" or growth terms and the "diffusion" or dispersal terms. The growth terms are usually similar to those introduced in the nineteenth century as models for populations without spatial heterogeneity. The simplest description of population dynamics is that a population will grow exponentially until limited by lack of available resources; this idea is generally credited to Thomas Malthus, who suggested it around 1800. If $P(t)$ denotes the population of some species at time t, then exponential growth is described by the equation

(1.1) $$P' = rP,$$

where r is the intrinsic growth rate of the population. A more sophisticated model, which explicitly incorporates the effects of limited resources on the population, is the logistic equation of Verhulst (1838, 1846). That equation is

(1.2) $$P' = r(1 - P/K)P,$$

where $K > 0$ denotes the carrying capacity of the environment. Equation (1.2) continues to be one of the most common starting points for modelling population dynamics. Related models for communities involving two or more species were proposed by Lotka (1925) and Volterra (1931); a typical model for a community with N species would be

$$(1.3) \qquad P'_\mu = (r_\mu + \sum_{\nu=1}^{N} a_{\mu\nu} P_\nu) P_\mu, \ \mu = 1, \cdots, N$$

where $a_{\mu\mu} < 0$ and the signs of the terms $a_{\mu\nu}$ with $\mu \neq \nu$ depend on the nature of the interactions between species. More general sorts of interactions which may be time dependent can also occur, and the corresponding extension of (1.3) is

$$(1.4) \qquad P'_\mu = f_\mu(t, P_1, \cdots, P_N) P_\mu$$

If the population being modelled can disperse through its environment, then the population density need not be uniform. Assuming that dispersal takes place via random walks or Brownian motion leads to a diffusion equation for the population density. When the diffusion is combined with dynamics such as those in (1.1)–(1.4), a reaction-diffusion model results. Such models were derived and studied by Skellam (1951) in the context of population dynamics. Skellam was inspired partly by the earlier work of Fisher (1930, 1937, 1950) on similar sorts of problems in population genetics. The most general sort of model considered by Skellam had the form

$$(1.5) \qquad u_t = d\Delta u + c_1(x)u - c_2(x)u^2,$$

where $u(x,t)$ represents the density of a population at time t and location x, d denotes the rate of diffusive dispersal, and $c_2(x) > 0$ describes the effects of limited resources. The intrinsic local growth rate, $c_1(x)$, could be positive or negative depending on local conditions. Of course, models more general than (1.5) may also arise; specifically, the diffusion process may be spatially variable or involve drift, the dynamics may be more complicated, or there may be time dependence in some of the coefficients. A fairly general model for a single species would be

$$(1.6) \qquad u_t = \nabla \cdot d(x,t) \nabla u - \vec{b}(x,t) \cdot \nabla u + f(x,t,u)u,$$

and a community could be modelled by adding diffusion terms to the sort of dynamics shown in (1.3) and (1.4). Here we shall focus our attention mainly on models of the forms (1.5) or (1.6), but without time dependence in the coefficients. (Some problems with periodic coefficients can be treated by the methods we shall describe, but in general a different approach is needed.) For further discussion of the modelling process, see (Skellam 1951; Okubo 1980; Levin 1986). The construction and analysis of reaction-diffusion models in ecology continues to be a very active and fruitful area of research in applied mathematics.

The connection between elliptic spectral theory and the sorts of ecological models we have described is shown clearly in the work of Skellam (1951). Skellam used classical methods to analyze a number of specific models. One typical example is the case of Malthusian growth on a circular region with a completely hostile exterior. The model is

(1.7)
$$u_t = d\Delta u + cu \text{ in } B_r \times (0, \infty)$$
$$u = 0 \text{ on } \partial B_r \times (0, \infty), \ u = u_0(x) \text{ on } B_r \times \{0\}$$

where $B_r = \{(x,y) : x^2 + y^2 < r^2\}$ and $u_0(x) \geq 0$, $u_0(x) \not\equiv 0$. Skellam used separation of variables to solve (1.7), and thus showed that the population modelled by (1.7) would grow if $c - (dj_1^2/r^2) > 0$ and decline if $c - (dj_1^2/r^2) < 0$. (Here, j_1 denotes the first zero of the Bessel function $J_0(s)$.) The connection between this result and spectral theory is of course that $(dj_1^2/r^2) - c$ is the first eigenvalue for $-d\Delta - c$ on B_r with homogeneous Dirichlet boundary conditions. The biological interpretation of the conclusion is that for the population to increase, the local growth rate as measured by c and the size of the domain as measured by r must be sufficiently large relative to the diffusion rate as measured by d. Such a conclusion is consistent with observations of natural populations; see, e.g., (Okubo 1980; Levin 1986). Another model considered by Skellam was the case of logistic growth on an interval. After suitable rescaling, that model becomes

(1.8)
$$u_t = u_{xx} + (1-u)u \text{ on } (-\ell, \ell) \times (0, \infty)$$
$$u(\pm \ell, t) = 0, \ u(x, 0) = u_0(x).$$

For ℓ sufficiently large, this model possesses a positive steady state which can be expressed in terms of elliptic functions. If u is such a steady state, it is said to be linearly stable if the spectrum of the linearized problem

(1.9)
$$\phi_{xx} + (1-2u)\phi + \sigma \phi = 0 \quad \text{on } (-\ell, \ell)$$
$$\phi(\pm \ell) = 0$$

lies in some set $\{z \in \mathbf{C} : \text{Re} z \geq \sigma_0 > 0\}$. In fact, classical Sturm-Liouville theory implies that the spectrum of (1.9) consists of a discrete set of eigenvalues $\sigma_1 < \sigma_2 < \sigma_3 < \cdots$, and the eigenfunction ϕ_1 corresponding to σ_1 does not change sign on $(-\ell, \ell)$. If we multiply (1.9) by u, integrate by parts, and use the fact that u is a steady state solution of (1.8) with $u > 0$ on $(-\ell, \ell)$, then we can solve for σ_1. We obtain

(1.10)
$$\sigma_1 = \int_{-\ell}^{\ell} u^2 \phi_1 \, dx \bigg/ \int_{-\ell}^{\ell} u\phi_1 \, dx > 0,$$

so that the steady state satisfies the criterion for linearized stability. Skellam considered several other models which could be analyzed via the classical methods of separation of variables, Sturm-Liouville theory, special functions, and so on. Of the more general model (1.5) he wrote (Skellam 1951, p.212) "Orthodox analytical methods appear inadequate, even in one-dimensional or radially symmetrical cases."

Since 1951 there have been great advances in analytical methods for partial differential equations, and there have been some changes in the way that the analysis is usually approached. Improvements in the theory of linear partial differential operators have made it possible to use powerful methods from functional analysis, and functional analysis has grown to encompass various results on nonlinear operators which permit the application of topological ideas to partial differential equations. These developments have affected the sorts of questions that are usually asked, shifting the focus from the quantitative to the qualitative. Such a shift is probably reasonable from the viewpoint of applications in ecology. Most real biological communities are far too complex to permit a detailed quantitative analysis. However, questions such as " How does the geometry of the habitat affect the population?" or "Can the existence of predators play a role in maintaining diversity in a community?" are at least partly qualitative, and are of considerable biological interest. Some of the major mathematical developments which are relevant to the study of such questions are modern bifurcation theory, the extension of the theory of dynamical systems to the infinite dimensional case, and the modern theory of positive operators, monotone iteration, and monotone flows. For bifurcation theory, a good general reference is (Chow and Hale 1982); the results in (Crandall and Rabinowitz 1971; Rabinowitz 1973) are widely considered to be especially important. For the application of the theory of dynamical systems to reaction-diffusion equations, a good source is (Henry 1981). The theory of positive operators is discussed at length in (Amann 1976), while ideas related to monotonicity are combined with the viewpoint of dynamical systems in (Matano 1984; Hirsch 1988; Smith 1988). A general overview of a number of modern methods for studying reaction-diffusion equations is given in (Smoller 1983). At the present time, our analytical methods are still inadequate for a complete understanding of models as general as (1.6) or for systems such as (1.4) with diffusive terms added; they are, however, sufficiently improved from those available in 1951 that a rigorous qualitative analysis of models such as (1.5) is now within our grasp, and we can say some things about more complicated problems.

2 The variational case

The model (1.5) has two features which make the analysis easier than in the more general case (1.6). The first is that the right side of (1.5) does not explicitly involve the time variable, so that (1.5) can be viewed as a dynamical system. The second is that the right side of (1.5) and the corresponding linearized operator are in variational or self-adjoint form, so that ideas such as the variational characterization of eigenvalues or integration by parts via the divergence theorem can be used efficiently. In fact, those two features taken together allow us to take the first step in our analysis by insuring that the dynamics of (1.5) are determined by its steady states. To be more specific, let us consider (1.5) for $(x,t) \in \Omega \times (0,\infty)$ with the boundary condition $u(x,t) = 0$ for $x \in \partial\Omega$, where Ω is a bounded domain in $\mathbf{R}^n$, $n = 1, 2$, or 3, where $\partial\Omega$ is smooth, and where c_1 and c_2 are at worst bounded measurable functions with $c_2(x) \geq c_2 > 0$. (The homogeneous Dirichlet boundary condition corresponds to a completely hostile exterior, such that any member of the population which reaches the boundary of Ω dies. If the exterior is hostile but not completely deadly, a mixed or Robin boundary condition results, and the analysis is similar. If the boundary acts as a barrier, so that individuals reaching the boundary of Ω simply return to the interior, a Neumann boundary condition results. The analysis may be somewhat different, since $-\Delta$ will have zero as an eigenvalue, but the same general approach can sometimes still be used. For a well motivated discussion of these boundary conditions from the modelling viewpoint, see (Ludwig, Aronson and Weinberger 1979).) When (1.5) is supplemented with appropriate boundary conditions, we may follow the ideas of (Henry 1981) and view it as generating a dynamical system on the Sobolev space $W^{1,2}(\Omega)$. Also, comparison theorems based on the maximum principle guarantee that nonnegative initial data remain nonnegative, and that orbits exist globally and remain bounded. Further, the dynamical system is of gradient type with corresponding Lyapunov functional

$$V(w) = \int_\Omega [d|\nabla w|^2/2 - c_1(x)w^2/2 + c_2(x)w^3/3]\,dx,$$

so it follows as in (Henry 1981, section 5.3) that the ω-limit set of (1.5) consists only of equilibrium points. Thus, if we can find and classify the nonnegative equilibrium states of (1.5), we should achieve a good understanding of its dynamics.

There are a number of approaches to analyzing the steady state problem for (1.5). One effective method is bifurcation theory. If we let $\lambda = 1/d$ in (1.5) and use the facts that with Dirichlet boundary conditions, $-\Delta$ is invertible with $(-\Delta)^{-1} : L^p(\Omega) \to W^{2,p}(\Omega) \cap W_0^{1,p}(\Omega)$ for any p, and that $W^{2,p}(\Omega) \cap W_0^{1,p}(\Omega)$ embeds compactly in $C_0^{1,\alpha}(\bar{\Omega})$ for any $\alpha < 1$, we can rewrite (1.5) in the form

(2.1) $$u = \lambda(-\Delta^{-1})c_1(x)u + \lambda\Delta^{-1}c_2(x)u^2.$$

The form of (2.1) is such that bifurcation theory can be applied. Observe that $u \equiv 0$ is always a solution, so that we may seek solutions bifurcating from the trivial solution branch $(\lambda, 0) \subseteq \mathbf{R} \times C_0^0(\bar{\Omega})$. (We could also work in $\mathbf{R} \times W_0^{1,p}(\Omega)$ or other spaces; our solutions will lie in $W^{2,p}(\Omega) \cap W_0^{1,p}(\Omega)$ for any p by the form of (2.1) and standard elliptic *a priori* estimates, such as those discussed in (Gilbarg and Trudinger 1977). The analysis of the mapping properties of elliptic operators is one of the improvements in the theory of linear differential operators which were mentioned earlier.) When we apply the results of (Crandall and Rabinowitz 1971) to (2.1), we may conclude that if the problem

(2.2) $$-\Delta\phi = \lambda c_1(x)\phi \text{ on } \Omega, \ \phi = 0 \text{ on } \partial\Omega$$

has a simple eigenvalue $\lambda_1 > 0$ with positive eigenfunction, then a branch of positive solutions (λ, u) bifurcates from the trivial solution branch at $\lambda = \lambda_1$. If we then apply the global bifurcation results of (Rabinowitz 1973), we may conclude that if no other positive eigenvalue of (2.2) has a positive eigenfunction, then the branch of positive solutions extends to infinity in either λ or $\|u\| \equiv \sup |u|$. Finally, we may use the maximum principle for weak solutions (Gilbarg and Trudinger 1977, Ch. 8) to conclude that all positive solutions satisfy $\|u\| \leq \text{ess sup } c_1(x)/\text{ess inf } c_2(x)$, so that the branch of positive solutions must meet infinity in λ. Since there are no nontrivial solutions when $\lambda = 0$, the branch must persist as $\lambda \to +\infty$, which is equivalent to $d \downarrow 0$. Thus, the question of existence of positive solutions to (2.1) and hence positive steady states for (1.5) can be answered if we know enough about the eigenvalues and eigenfunctions of (2.2). For a more detailed analysis of this sort of problem, see (Cantrell and Cosner, preprint b).

In problem (2.2) if $c_1(x) > 0$ in Ω, then the eigenvalues and eigenfunctions can be analyzed via classical variational principles (Courant and Hilbert 1953). However, we wish to consider cases where $c_1(x)$ may change sign in Ω. (Recall that (1.5) is a model for a population inhabiting a spatially heterogeneous environment, and that $c_1(x)$ represents the intrinsic growth rate at the point x; hence $c_1(x) > 0$ in regions of favourable habitat and $c_1(x) < 0$ in unfavourable regions.) It turns out that the methods of (Courant and Hilbert 1953) extend fairly directly to problems such as (2.2) with weights which are indefinite, i.e., which change sign in Ω. Some examples are discussed in (Weinberger 1974), and a general theorem is stated in (Manes and Micheletti 1973):

Theorem 2.1 *Suppose that* $c_1(x) \in L^p(\Omega)$, *with* $p \geq n/2$, *and that the sets* $\{x \in \Omega : c_1(x) > 0\}$ *and* $\{x \in \Omega : c_1(x) < 0\}$ *have positive measure. Then the problem (2.2) has a doubly infinite sequence of eigenvalues*

$$\cdots \lambda_n^- \leq \cdots \lambda_3^- \leq \lambda_2^- < \lambda_1^- < 0 < \lambda_1^+ < \lambda_2^+ \leq \lambda_3^+ \leq \cdots \leq \lambda_n^+ \leq \cdots,$$

with λ_1^- *and* λ_1^+ *simple and the only eigenvalues which admit positive eigenfunctions. The eigenvalues can be characterized variationally by*

$$(2.3) \quad \begin{aligned} (\lambda_n^+)^{-1} &= \sup\inf_{F_n} \{\int_\Omega c_1(x)\phi^2 \, dx : \phi \in F_n, \quad \int_\Omega |\nabla \phi|^2 dx = 1\} \\ (\lambda_n^-)^{-1} &= \inf\sup_{F_n} \{\int_\Omega c_1(x)\phi^2 \, dx : \phi \in F_n, \quad \int_\Omega |\nabla \phi|^2 dx = 1\} \end{aligned}$$

where F_n *ranges over all n-dimensional subspaces of* $W_0^{1,2}(\Omega)$. *The eigenvalues depend continuously on* $c_1(x)$ *in the topology of* $L^{n/2}(\Omega)$ *and are nondecreasing with respect to* $c_1(x)$.

Remarks: Manes and Michletti (1973) consider more general second order elliptic operators in divergence form, and obtain results on the continuity of the eigenvalues with respect to the coefficients of the operator. It should be possible to treat the case of classical Robin (i.e., third type or mixed) boundary conditions in a similar way. The case of pure Neumann boundary conditions is somewhat different and will be discussed later. Some applications of results such as Theorem 2.1 to nonlinear problems are given in (Manes and Micheletti 1973; de Figuieredo 1982). Methods of numerical approximation of eigenvalues are discussed in (Weinberger 1974).

When combined with the analysis of the preceding paragraphs, Theorem 2.1 implies that the problem (1.5) with homogeneous Dirichlet boundary conditions has a positive steady state solution for $0 < d < 1/\lambda_1^+$. There remains a number of questions, however. Are there any positive steady states for $d \geq 1/\lambda_1^+$? When there is a positive steady state, is it unique, and is it stable? How does the value of λ_1^+ depend on $c_1(x)$? It turns out that variational methods can be used to help answer these questions, but that raises another question, namely, what if we consider problems with drift terms or systems, which generally will not have a variational formulation? We shall return to the last question later; for the moment, we shall continue to explore the implications of the variational theory.

If we apply (2.3) in the case of λ_1^+, it follows that for any $\phi \in W_0^{1,2}(\Omega)$ we have $\int_\Omega c_1 \phi^2 dx \leq (1/\lambda_1^+) \int_\Omega |\nabla \phi|^2 dx$. If we have a positive solution $u \in W^{2,p}(\Omega) \cap W_0^{1,p}(\Omega)$ for

(2.4) $$0 = d\Delta u + c_1(x)u - c_2(x)u^2 \text{ in } \Omega,$$

then multiplying by u and integrating by parts yields

$$d\int_\Omega |\nabla u|^2 dx + \int_\Omega c_2 u^3 dx = \int_\Omega c_1 u^2 dx \leq (1/\lambda_1^+) \int_\Omega |\nabla u|^2 dx.$$

Since $c_2(x) > 0$ in Ω, it follows that $d < 1/\lambda_1^+$, so that the condition is necessary as well as sufficient for the existence of positive equilibrium states.

After we have established conditions for the existence of positive steady states for (1.5), the next question is to determine the multiplicity and properties, especially stability or instability, of steady states. It turns out that linear spectral theory can be used to address both the question of existence and that of stability. The ideas are taken from the work of Hess (1977, 1982a); related results are used in (Blat and Brown 1984; Cantrell and Cosner 1987; Cantrell and Cosner, preprint a). The relevant spectral result is the following theorem, which can be established via the variational methods of (Courant and Hilbert 1953; Weinberger 1974).

Theorem 2.2 *Suppose that $q(x) \in L^\infty(\Omega)$ and $d > 0$. The problem*

(2.5) $$-\Delta\psi + q(x)\psi = \sigma\psi \text{ on } \Omega, \ \psi = 0 \text{ on } \partial\Omega$$

has an infinite sequence of eigenvalues $\sigma_1 < \sigma_2 \leq \sigma_3 \leq \cdots$. The eigenvalue σ_1 is simple and is the only eigenvalue admitting a positive eigenfunction. It is strictly increasing with respect to q in the sense that if $q_1(x) < q_2(x)$ on Ω, then the corresponding principal eigenvalues satisfy $\sigma_1(q_1) < \sigma_1(q_2)$.

Remark: The differential operator in (2.5) is a Schrödinger operator, and spectral problems such as (2.5) have been widely studied by mathematical physicists. (Problems such as (2.2) occur less frequently in mathematical physics and have received somewhat less attention.) The spectral theory of Schrödinger operators has a vast literature; a good introduction is (Reed and Simon 1978).

To obtain a uniqueness result for solutions of (2.4) from Theorem 2.2, suppose that u_1 and u_2 are two positive solutions to (2.4) with $u_1 = u_2 = 0$ on $\partial\Omega$. Then the eigenvalue problem of form (2.5) given by

$$(2.6) \qquad -d\Delta\psi + [c_2(x)u_1 - c_1(x)]\psi = \sigma\psi \text{ on } \Omega, \ \psi = 0 \text{ on } \partial\Omega$$

has principal eigenvalue $\sigma_1 = 0$, since u_1 is an eigenfunction, $u_1 > 0$ by assumption, and σ_1 is the only eigenvalue with positive eigenfunction. Let $w = u_1 - u_2$; then $w = 0$ on $\partial\Omega$, and

$$(2.7) \qquad -d\Delta w + [c_2(x)(u_1 + u_2) - c_1(x)]w = 0 \text{ on } \Omega.$$

But $c_2(x) > 0$ and $u_2 > 0$, so the principal eigenvalue, and hence all the eigenvalues, of

$$(2.8) \qquad -d\Delta\psi + [c_2(x)(u_1 + u_2) - c_1(x)]\psi = \sigma\psi \text{ on } \Omega, \ \psi = 0 \text{ on } \partial\Omega$$

must be strictly greater than the principal eigenvalue of (2.6) which is zero. Thus, zero is not an eigenvalue of (2.8), so (2.7) cannot have any solutions other than $w \equiv 0$, and it follows that $u_1 \equiv u_2$. Clearly, this analysis applies to problems more general than (2.4); in particular, it extends to formally self-adjoint uniformly elliptic problems of the form

$$\sum_{i,j=1}^{n}(a_{ij}(x)u_{x_i})_{x_j} + f(x,u)u = 0 \text{ in } \Omega, u = 0 \text{ on } \partial\Omega,$$

provided that $f(x,t) \leq 0$ for $t \geq u_0 > 0$ and that $f(x,t)$ is strictly decreasing in t for $t \in [0, u_0]$ (Hess 1977).

The problem of stability can be treated in an analogous way. If $u_1 > 0$ is a solution to (2.4), then linearization of the problem about u yields the operator $-d\Delta + 2c_2(x)u_1 - c_1(x)$. Again, comparison with (2.6) via Theorem 2.2 shows that the principal eigenvalue of the linearized operator is positive, which is exactly what is needed to establish linearized stability. The stability result can be sharpened by using the principal eigenfunction to construct suitable upper and lower solutions and then using a comparison theorem for (1.5) based on the maximum principle. This idea is discussed in (Hess 1982a) and is used to study problems related to (1.5) in (Cantrell and Cosner, preprints a,b).

As noted earlier, the case of Neumann boundary conditions differs somewhat from those of Dirichlet or Robin conditions. The main difference is that in the problem corresponding to (2.2) with the Neumann condition $\partial\phi/\partial\nu = 0$ on $\partial\Omega$ (where $\partial/\partial\nu$ denotes a normal derivative), zero is always an eigenvalue with constant eigenfunction, and there may or may not be a positive eigenvalue with positive eigenfunction. The relevant result is given in (Brown and Lin 1980):

Theorem 2.3 *Suppose that $c_1(x) \in L^\infty(\Omega)$ and that the sets $\{x \in \Omega : c_1(x) > 0\}$ and $\{x \in \Omega : c_1(x) < 0\}$ both have positive measure. If $\int_\Omega c_1(x)dx < 0$, then the* *problem*

(2.9) $\qquad -\Delta\phi = \mu c_1(x)\phi$ *on* $\Omega, \partial\phi/\partial\nu = 0$ *on* $\partial\Omega$

has a unique positive eigenvalue μ_1 admitting a positive eigenfunction, which can be characterized by

(2.10)
$$\mu_1 = \inf\{\int_\Omega |\nabla\phi|^2 dx / \int_\Omega c_1(x)\phi^2 dx : \phi \in W^{2,2}(\Omega),$$
$$\partial\phi/\partial\nu = 0 \text{ on } \partial\Omega, \text{ and } \int_\Omega c_1(x)\phi^2 dx > 0\}.$$

If $\int_\Omega c_1(x)dx \geq 0$, then there is no positive eigenvalue whose eigenfunction does not vanish in Ω.

Remark: The characterization (2.10) is equivalent to that given in formula (2.3) of Theorem 2.1 for λ_1^+, except for the space from which ϕ may be taken.

The case of Neumann conditions is important in applications for two reasons. It occurs naturally in models for population dynamics where the boundary of the habitat is not deadly but simply impassable, and it is the natural boundary condition for certain selection-migration models in population genetics. The equations arising in population genetics typically have the form

$$u_t = d\Delta u + g(x)h(u) \text{ in } \Omega \times (0,\infty)$$

$$\partial u/\partial\nu = 0 \text{ on } \partial\Omega \times (0,\infty), u = u_0(x) \geq 0 \text{ on } \Omega \times \{0\},$$

where u represents the frequency of occurrence of some allele in a population, so $0 \leq u \leq 1$, and where $h(0) = h(1) = 0$. Such equations were considered by Fisher (1930, 1937, 1950) and have been studied by various investigators; see (Fleming 1975; Fife and Peletier 1977; Saut and Scheurer 1978; Brown and Lin 1980, 1981). In the context of population dynamics, the behaviour of (1.5) subject to homogeneous Neumann conditions is similar to that under Dirichlet conditions if $\int_\Omega c_1(x)dx < 0$ so that there is a positive eigenvalue with positive eigenfunction to act as a bifurcation point for positive steady states. If $\int_\Omega c_1(x)dx \geq 0$, then there will exist positive steady states for all values of d; that is the main qualitative difference. A biological interpretation is that when the environment has an impassable boundary and is on the average unfavourable ($\int_\Omega c_1(x)dx < 0$) then high diffusion rates have the same effect (namely, the ultimate extinction of the population) as they *always* have when the boundary is deadly; but if the boundary is impassable and the environment is on the average neutral or favourable ($\int_\Omega c_1(x)dx \geq 0$) then the population can persist, no matter what its rate of diffusion.

If we consider only the case of Dirichlet conditions, i.e., the case where the boundary of the habitat is deadly to the population, the spatial distribution of positive and negative regions for $c_1(x)$ can still have a significant quantitative effect on the behaviour of (1.5). The condition which is necessary and sufficient for the existence of a unique positive, stable steady state for (1.5) is $d < 1/\lambda_1^+(c_1(x))$; if this condition is met, (1.5) predicts persistence for the population, and otherwise (1.5) predicts extinction. Thus, increasing $\lambda_1^+(c_1(x))$ imposes a more stringent condition on the diffusion rate if the population is to persist. The next two results suggest two factors which affect $\lambda_1^+(c_1(x))$.

Theorem 2.4 *Suppose that for $j = 1, 2, 3, \cdots$, $m_j(x) \in L^\infty(\Omega)$ with $\|m_j\|_\infty \leq M_0$, and $m_j(x) > 0$ on a set of positive measure. If $\lambda_1^+(m_j(x))$ is the positive principal eigenvalue for the problem (2.2) with $c_1(x)$ replaced by $m_j(x)$, then a necessary and sufficient condition for having $\lim_{j \to \infty} \lambda_1^+(m_j(x)) = \infty$ is that*

$$\lim_{j \to \infty} \sup \int_\Omega m_j(x) \psi(x) dx \leq 0 \text{ for all } \psi \in L^1(\Omega) \text{ with } \psi \geq 0 \text{ a.e.}$$

Theorem 2.4 is proved in (Cantrell and Cosner, preprint b). The proof is based on the variational formulation (2.3) for $\lambda_1^+(m_j(x))$. The next result was obtained by R. S. Cantrell via a direct computation:

Theorem 2.5 *Suppose that $\Omega = (-1, 1) \subseteq \mathbf{R}^1$ and $c_1(x) = 1$ on $(x_0 - \delta, x_0 + \delta) \subseteq (-1, 1)$ and $c_1(x) = -1$ on $(-1, x_0 - \delta) \cup (x_0 + \delta, 1)$. Then $\lambda_1^+(c_1(x))$ increases monotonically as x_0 moves away from 0 toward ± 1.*

Theorem 2.4 and 2.5 and some related results and numerical experiments indicate that both the extent to which favourable and unfavourable habitats are intermingled and the proximity of the favourable regions to the deadly boundary have an effect on the suitability of the environment for a population modelled by (1.5). Theorem 2.4 and the Riemann-Lebesgue lemma imply that if $\Omega = (-1, 1)$ and $c_{1j}(x) = \sin(jx)$, then $\lim_{j \to \infty} \lambda_1^+(c_{1j}(x)) = \infty$; so in a sense, the theorem says that for highly oscillatory growth rates, corresponding to closely intermingled regions of favourable and unfavourable habitat, the overall suitability of the environment will be low. However, Theorem 2.5 shows that not only the size and number of good and bad regions affect the overall quality of the environment, but so does their arrangement relative to the boundary. The question of how $\lambda_1^+(c_1(x))$ depends on $c_1(x)$ and the corresponding questions for more general operators seem to be fairly complicated. They have not yet been completely answered.

The variational approach to spectral theory has many advantages, but also some limitations. It does not generally apply directly to problems involving drift as well as diffusion, i.e., to problems involving elliptic operators which are not formally self-adjoint. It does not apply to many systems. In the next section, we shall explore an alternative approach.

3 Nonself-adjoint problems

It is natural to ask whether the criteria for persistence or extinction for populations modelled by (1.5) have any reasonable analogues for more complicated situations described by (1.6). The analysis of fully time dependent models such as (1.6) obviously cannot be based entirely on the examination of the steady states, since there generally will not be any except $u \equiv 0$, but most of the results of the last section do extend to models of the form

(3.1)
$$u_t = \nabla \cdot d(x) \nabla u - \vec{b}(x) \cdot \nabla u + c_1(x)u - c_2(x)u^2 \text{ in } \Omega \times (0, \infty)$$
$$u = 0 \text{ on } \partial\Omega \times (0, \infty), u = u_0 \geq 0 \text{ on } \Omega \times \{0\}.$$

It is somewhat surprising that many of the results obtained for (1.5) extend to (3.1), since the variational methods used to derive them break down. The key idea is to replace arguments based on self-adjointness or variational structure with others based on positivity or monotonicity. This idea has been used for quite a while in matrix theory. Any real, symmetric matrix will have real eigenvalues, while an arbitrary square matrix may have only complex eigenvalues. However, if a square matrix has all of its entries positive, the Perron-Frobenius theorem implies that it must have at least one real eigenvalue, and in fact that eigenvalue is positive, has the greatest absolute value of any eigenvalue, and admits a componentwise positive eigenvector. The spectral theory needed to analyze (3.1) can be derived via arguments based on the infinite dimensional generalizations of the Perron-Frobenius theorem, such as the Krein-Rutman theorem.

Before we attempt to analyze the steady states of (3.1), we must ask whether those steady states determine the dynamics of (3.1). The equation still induces a semiflow on appropriate Sobolev spaces, but it is not of gradient type, so the results of (Henry 1981) cannot be used. However, the semiflow is monotone in the sense that if u and v are two solutions to (3.1) with $u(x,0) \geq v(x,0)$ and $u(x,0) \not\equiv v(x,0)$ for $x \in \Omega$, then $u > v$ on $\Omega \times (0, \infty)$ and $\partial u/\partial \nu < \partial v/\partial \nu$ on $\partial\Omega \times (0, \infty)$, where $\partial/\partial\nu$ is the outer normal derivative on $\partial\Omega$. This monotonicity property follows from comparison theorems based on the maximum principle; see (Smoller 1983) for a discussion of such results. It permits the application of recent results on monotone semiflows derived in (Matano 1984; Hirsch 1988). A historical overview and survey of the theory of monotone dynamical systems is given in (Smith 1988); here we shall only describe the implications of that theory for (3.1). There are two sorts of results which are relevant, First, almost all orbits which are globally bounded in $L^\infty(\Omega)$ with respect to t are attracted to the set of equilibria, so the steady states generically at least determine the dynamics. (Here, boundedness in $L^\infty(\Omega)$ and regularity theory imply compactness of the closures of orbits, and "almost all" can be interpreted in terms of cardinality, measure, or topology; see (Hirsch 1988).) Second, if there are two steady states v and w with $v > w$ on Ω and no steady states between v and w, then either all solutions u of (3.1) with initial data u_0 satisfying $v \geq u_0 \geq w$ on Ω, $u_0 \not\equiv w$, converge to v, or all solutions with initial data satisfying $v \geq u_0 \geq w$ on Ω, $u_0 \not\equiv v$, converge to w. Thus, if we can establish conditions under which (3.1) has a unique positive linearly stable steady state, then by taking v to be that steady state and $w \equiv 0$ we may conclude that the population density will tend to v, implying persistence; if there is no such steady state, then

(3.1) predicts that almost all initial distributions of the population will ultimately decline to extinction. Hence, the answer to the question "do the steady states of (3.1) determine the dynamics?" is "generically, yes they do."

To analyze the steady states of (3.1) via bifurcation theory, it is convenient to introduce a parameter λ to "unfold" the steady state equation into the form

(3.2)
$$0 = \nabla \cdot d(x) \nabla u - \vec{b}(x) \cdot \nabla u + \lambda[c_1(x)u - c_2(x)u^2] \text{ in } \Omega,$$
$$u = 0 \text{ on } \partial\Omega.$$

If we can establish conditions on λ which imply or preclude the existence of positive solutions to (3.2), we can then check whether $\lambda = 1$ satisfies those conditions and thus draw conclusions about the steady states of (3.1). The key issue then becomes whether (3.2) has any bifurcation points for positive solutions. If so, then the same sort of analysis used to establish conditions for the existence of positive steady states for (1.5) can be extended to (3.2). The bifurcation theoretic results of (Crandall and Rabinowitz 1971; Rabinowitz 1973) are stated abstractly, and standard elliptic theory as discussed in (Gilbarg and Trudinger 1977) implies that the elliptic operator $-\nabla \cdot d(x) \nabla + \vec{b}(x) \cdot \nabla$ has a compact inverse on a number of suitable function spaces, so that (3.2) can be rewritten in a form analogous to (2.1); the only missing ingredient is the linear elliptic spectral theory.

To obtain a positive solution of (3.2) from bifurcation theory, we need to find a positive real eigenvalue for the linearization of (3.2) about the trivial solution $u \equiv 0$ which admits a positive eigenfunction. The existence of such an eigenvalue is not obvious, since the elliptic operator in (3.2) is not in self-adjoint form, and may thus have complex spectrum. However, it turns out that under reasonable hypotheses such an eigenvalue will always exist. The existence of a positive simple eigenvalue with positive eigenfunction was shown for general second order elliptic operators in the case of a positive weight function by Protter and Weinberger (1966). Their idea was to observe that if L is a second order elliptic operator with fairly smooth coefficients having the form

$$Lu = -\sum_{i,j=1}^{n} a_{ij}(x)u_{x_i x_j} + \sum_{i=1}^{n} b_i(x)u_{x_i},$$

then under homogeneous Dirichlet boundary conditions, L^{-1} can be extended to a compact operator from $C_0^0(\bar{\Omega})$ into itself. If $c(x) > 0$, then the problem $L\phi = \lambda c(x)\phi$ on Ω, $\phi = 0$ on $\partial\Omega$, can be rewritten as $\phi = \lambda L^{-1}c(x)\phi$, and by the maximum principle, $L^{-1}[c(x)f] > 0$ in Ω if $f \geq 0$ in Ω, $f \not\equiv 0$. This combination of positivity and compactness permits the application of the Krein-Rutman theorem, which states that a compact, strongly positive operator T with positive spectral radius $r(T)$ defined on an ordered Banach space whose positive cone has nonempty interior has a unique, simple positive eigenvalue $\sigma_1 = r(T)$ with positive eigenvector, and for positive f, the problem $\sigma u - Tu = f$ has a unique positive solution u if $\sigma > \sigma_1$ and no positive solution if $\sigma \leq \sigma_1$. (For a detailed discussion of the Krein-Rutman theorem and other results for ordered Banach spaces, see (Amann 1976).) The present object is just to give a rough description of the abstract results used to develop the parts of the spectral theory for nonself-adjoint operators that are needed for our analysis of (3.2).) If we take $\lambda_1 = 1/\sigma_1$, then λ_1 is the positive eigenvalue we need for the problem $L\phi = \lambda c(x)\phi$ on Ω, $\phi = 0$ on $\partial\Omega$; also, the operator $L - \lambda c(x)$ will have a positive inverse for $\lambda < \lambda_1$ and will admit no positive solutions to $Lu - \lambda c(x)u = f > 0$ for $\lambda \geq \lambda_1$. The results of (Protter and Weinberger 1966) also include some estimates for λ_1, which we shall discuss later, and the additional result that for *any* eigenvalue λ for $L\phi = \lambda c(x)\phi$ on Ω, $\phi = 0$ on $\partial\Omega$, we have $\text{Re}\lambda \geq \lambda_1$. For our purpose, we must consider the case where $c_1(x)$ may change sign. Results similar to those of (Protter and Weinberger 1966) were obtained for that case by (Hess and Kato 1980). The arguments again depend on the theory of positive operators, but because of the indefiniteness of $c_1(x)$ the analysis is more complicated. Some refinements of the results of (Hess and Kato 1980) are given in (Gossez and Lami Dozo 1982, 1985; Hess 1982b; Nussbaum 1984). The ideas of (Protter and Weinberger 1966) were extended to parabolic problems with periodic coefficients by Lazer and his co-workers (Lazer 1982; Castro and Lazer 1982); the case of periodic parabolic problems with indefinite weights was considered in (Beltramo and Hess 1984; Hess 1984). The case of Neumann boundary conditions for an elliptic problem is discussed in (Senn and Hess 1982), where results analogous to those of Theorem 2.3 are extended to the nonself-adjoint case. Some results for systems of equations are given in (Amann 1976); for systems with indefinite weights see (Hess 1983; Cantrell and Schmitt 1986). The following is a corollary of the results of (Hess and Kato 1980):

Theorem 3.1 *Suppose that $d(x) \in C^{1,\alpha}(\bar{\Omega})$, $\vec{b}(x) \in [C^\alpha(\bar{\Omega})]^n$ and $c_1(x) \in C(\bar{\Omega})$, with $d(x) \geq d_0 > 0$ for some constant d_0 and with $c_1(x_0) > 0$ for some $x_0 \in \Omega$. The problem*

$$(3.3) \qquad -\nabla \cdot d(x)\nabla\phi + \vec{b}(x) \cdot \nabla\phi = \lambda c_1(x)\phi \text{ on } \Omega, \phi = 0 \text{ on } \partial\Omega,$$

has a unique positive eigenvalue $\lambda_1^+(d,\vec{b},c_1)$ with positive eigenfunction. If λ is any other eigenvalue with $\text{Re}\lambda \geq 0$, then $\text{Re}\lambda \geq \lambda_1^+(d,\vec{b},c_1)$. If $c_1(x) > \tilde{c}_1(x)$ in Ω, then $\lambda_1^+(d,\vec{b},c_1) < \lambda_1^+(d,\vec{b},\tilde{c}_1)$. If $h(x) \in C(\bar{\Omega})$, $h(x) \geq 0$ on Ω and $h \not\equiv 0$, and v satisfies

$$(3.4) \qquad -\nabla \cdot d(x)\nabla v + \vec{b}(x) \cdot \nabla v - \lambda c_1(x)v = h(x) \text{ in } \Omega, v = 0 \text{ on } \partial\Omega,$$

then for $0 \leq \lambda < \lambda_1^+(d,\vec{b},c_1)$, we have $v > 0$ in Ω, while if $\lambda \geq \lambda_1(d,\vec{b},c_1)$ then $v \not> 0$ in Ω. The value $\sigma_1 = 1/\lambda_1^+(d,\vec{b},c_1)$ is an eigenvalue of algebraic multiplicity 1 for the operator $Tw \equiv [-\nabla \cdot d\nabla + \vec{b} \cdot \nabla]^{-1}c_1(x)w$ on $C_0^0(\bar{\Omega})$. If $c_1(x_1) < 0$ for some other $x_1 \in \Omega$, there exists a corresponding negative eigenvalue $\lambda_1^-(d,\vec{b},c_1)$.

Remark: The condition that $c_1(x) \in C(\bar{\Omega})$ is not essential provided that $c_1(x) \in L^\infty(\Omega)$ and we can construct a test function having certain properties. If $c_1(x) \geq$

$c_0 > 0$ on some open subset of Ω, then the arguments of Hess and Kato (1980) go through without any major modifications; the general case of $c_1(x) \in L^\infty(\Omega)$, $c_1(x) > 0$ on a set of positive measure, requires some other ideas and we shall not discuss it further here.

Theorem 3.1 is precisely the sort of result that is needed for the bifurcation theoretic analysis of (3.2). The conclusion of that analysis is that (3.2) has a positive solution provided $\lambda > \lambda_1^+(d, \vec{b}, c_1)$. Since the steady states of (3.1) correspond to the case $\lambda = 1$ in (3.2), we find that (3.1) has a positive steady state provided that $\lambda_1^+(d, \vec{b}, c_1) < 1$. Recall that for the case of (1.5), the condition for existence of a positive steady state was equivalent to $d\lambda_1^+(c_1(x)) < 1$, which is in turn a special case of the condition for (3.1). The monotonicity of $\lambda_1^+(d, \vec{b}, c_1)$ with respect to $c_1(x)$ and the positivity properties of (3.4) allow us to show uniqueness and stability of the positive steady state of (3.1) in much the same way as in the variational case; we omit the details.

Once we have extended the qualitative theory for (1.5) to (3.1), the next question is how $\lambda_1^+(d, \vec{b}, c_1)$ depends on d, $\vec{b}$, and c_1. A new question in this context is how $\vec{b}$ affects the eigenvalue. In some cases, it is possible to compare problems with drift to related self-adjoint problems and thus obtain some information. There are also some estimates which can be made directly from (3.3) provided appropriate comparison functions can be found. As noted in the previous section, the dependence of λ_1^+ on the coefficients is not completely understood even when d is constant and $\vec{b}$ is zero. In the general case there are many more questions that can be asked, and at the present time most of them remain open.

If we impose a coercivity condition on the differential operator in (3.3), we can extend Theorem 2.4 to that problem. Specifically, if there is a constant $d_0 > 0$ such that

$$(3.5) \quad \int_\Omega [d(x)|\nabla \phi|^2 + (\vec{b}(x) \cdot \nabla \phi)\phi] dx \geq d_0 \int_\Omega |\nabla \phi|^2 dx$$

for all $\phi \in W_0^{1,2}(\Omega)$ and if the weights $m_j(x)$ satisfy the hypotheses on $c_1(x)$ in Theorem 3.1 and are uniformly bounded in $L^\infty(\Omega)$, then $\lim_{j \to \infty} \lambda^+(d, \vec{b}, m_j) = \infty$ if and only if

$$\lim_{j \to \infty} \sup \int_\Omega m_j(x)\psi dx \leq 0 \text{ for all } \psi \in L^1(\Omega) \text{ with } \psi \geq 0 \text{ almost everywhere.}$$

The coercivity condition (3.5) will hold under various hypotheses, for example if we assume $d(x) \geq d_1 > 0$, $|\vec{b}| \leq b_1$, and $d_1 > b_1/\sqrt{\lambda_1^+(1,0,1)}$. The extension of Theorem 2.4 to the case of (3.3) is based on comparing quantities obtained from (3.3) via integration by parts to variational integrals of the sort discussed in the preceding section of this article. To show the necessity of the condition on the weight m_j, it is convenient to use an idea of Holland (1978) which gives a variational characterization of the principal eigenvalue for nonself-adjoint elliptic problems in the case of positive weights. A related argument is used in (Hess 1982b) to obtain upper bounds on the principal eigenvalue for a periodic parabolic problem with an indefinite weight, but the results of (Holland 1978) have not yet been extended fully to the case of sign indefinite weights.

If the drift term in (3.3) is viewed as a perturbation from a self-adjoint problem, it is sometimes possible to obtain bounds on λ_1^+ which give information about the qualitative effects of the drift term. This approach was taken by Murray and Sperb (1983), who showed that if $d \equiv 1$, $c_1 \equiv 1$, and $\vec{b} = -\nabla B$ with $\gamma_1 \leq (\Delta B/2) + (|\nabla B|^2/4) \leq \gamma_2$ then $\lambda_1^+(1,0,1) + \gamma_1 \leq \lambda_1^+(1,\vec{b},1) \leq \lambda_1^+(1,0,1) + \gamma_2$. A more general but less precise result is

Theorem 3.2 *Suppose that either $\vec{b} \in [C^1(\bar{\Omega})]^n$ with $\nabla \cdot \vec{b} \leq 0$ or that $\vec{b} = -d(x) \nabla B(x)$ for some $B(x) \in C^2(\bar{\Omega})$ with $(\nabla \cdot d \nabla B/2) + (d|\nabla B|^2/4) \geq 0$. Then $\lambda_1^+(d,\vec{b},c_1) \geq \lambda_1^+(d,0,c_1)$.*

Remarks: The results of (Murray and Sperb 1983) and of Theorem 3.2 imply that adding a constant drift term always raises the principal eigenvalue of a self-adjoint problem. However, it is possible to give examples where a variable drift term can lower that eigenvalue. The physical or biological interpretation is not entirely clear. The general question of comparing $\lambda_1^+(d,\vec{b}_1,c_1)$ with $\lambda_1^+(d,\vec{b}_2,c_1)$ is apparently open. There are many questions of this sort which merit further study.

There is another sort of estimate for λ_1^+ which does not depend on comparison with problems in variational form. This type of estimate was introduced for the Laplace operator with constant weight by Barta (1937) and extended to general second order elliptic problems with positive weights by Protter and Weinberger (1966). Barta's estimate essentially says that if $w > 0$ in Ω and $w \in C^2(\bar{\Omega})$, then $\lambda_1^+(1,0,1) \geq \inf(-\Delta w/w)$. Analogous results for the case of indefinite weights were given by Gossez and Lami Dozo (1982, 1985) and Nussbaum (1984); a version of the result in (Nussbaum 1984) for (3.3) is

Theorem 3.3 *Suppose that the hypotheses of Theorem 3.1 are satisfied with $c_1(x) \in C^\alpha(\bar{\Omega})$. Suppose also that $w \in C^{2,\alpha}(\bar{\Omega})$ with $w = 0$ on $\partial\Omega$, $w \geq 0$ on Ω, $w \not\equiv 0$, such that for some $\rho < \lambda_1^+(d,\vec{b},c_1)$, we have $-\nabla \cdot d \nabla w + \vec{b} \cdot \nabla w - \rho c_1 w \geq 0$ for all $x \in int\{y \in \Omega : c_1(y) = 0\}$. Let $\Lambda(w) = \sup\{\lambda : -\nabla \cdot d \nabla w + \vec{b} \cdot \nabla w - \lambda c_1 w \geq 0$ for all $x \in \Omega$ with $c_1(x) > 0\}$. Then $\lambda_1^+(d,\vec{b},c_1) > \Lambda(w)$ unless w is a positive multiple of the eigenfunction associated with $\lambda_1^+(d,\vec{b},c_1)$, in which case $\Lambda(w) = \lambda_1^+(d,\vec{b},c_1)$.*

Remarks: The difficulty in using a result such as Theorem 3.3 is in constructing an appropriate comparison function w. A technique for doing that in principle is given in (Gossez and Lami Dozo 1985), but the method may be tricky to use in a particular problem. A number of related ideas and results are discussed in (Gossez and Lami Dozo 1985; Nussbaum 1984); the latter has a fairly extensive list of references. The estimates of the sort given in Theorem 3.3 depend on the maximum principle, either directly or through the theory of positive operators.

As we have seen, there now exist analytical methods powerful enough for the analysis of Skellam's problem (1.5), and even more general problems. However, we do not yet have a complete understanding of the implications of those methods for problems of mathematical ecology. To obtain a more complete picture will require more refined results about the dependence of the principal eigenvalue on the coefficients of the problem, and more detailed interpretations of those results in the ecological context. Both of those areas should provide numerous opportunities for further research by mathematicians with a taste for applied problems or biologists with a taste for mathematics.

4 Some related ideas

The ideas described in the preceding sections of this article have numerous extensions, generalizations, and connections with other sorts of mathematical problems. There is a vast literature on elliptic spectral theory, and mathematical ecology is currently a very active area of applied mathematics. One natural generalization of the problems and methods we have described is to systems of equations modelling ecological communities. A typical system would have dynamics similar to those in (1.3) or (1.4) coupled with diffusion terms as in (1.5) or (1.6), e.g.,

$$(4.1) \qquad u_t^j = \nabla \cdot d(x) \nabla u^j - \vec{b} \cdot \nabla u^j + f^j(x, \vec{u}) u^j, j = 1, \cdots, N,$$

where u^j represents the population density of the jth species. Treatments of problems such as (4.1) based in part on spectral theory are given in (Blat and Brown 1984; Cantrell and Cosner 1987a), among other references. Some spectral results for systems which involve indefinite weights are given in (Hess 1983; Cantrell and Schmitt 1986), but the theory is not as well developed as for a single equation. Some other results about the spectra of elliptic systems which are motivated by questions about reaction-diffusion problems are discussed in (Brown and Eilbeck 1982; Cantrell and Cosner 1987b).

There are a number of results on elliptic eigenvalue problems with indefinite weights which do not have immediate applications to ecological models. Some results on the completeness of eigenfunction expansions are given in (Beals 1985; Hess 1985; Faierman and Roach 1987). The asymptotic distribution of eigenvalues is discussed in (Fleckinger and Lapidus 1986; Hess 1986). There has been a considerable amount of work on Sturm-Liouville problem with indefiniteness in both the differential operator and the weight function, as opposed to the cases we have considered where the indefiniteness is only in the weight. It turns out that a problem of the form $(p(x)y')' + q(x)y + \lambda r(x)y = 0$ on $(0, 1)$ with $y(0) = y(1) = 0$ and $p > 0$ can have multiple or even complex eigenvalues unless certain sign or definiteness conditions are imposed on q and r. For a discussion of these phenomena and some interesting historical references, see (Fleckinger and Mingarelli 1984; Mingarelli 1982, 1983). The richness of the mathematical phenomena and the connections with applied problems have kept elliptic spectral theory an important and lively area of research since the time of Fourier, and promise to maintain activity in the area for a long time to come. Perhaps the examples we have described will interest some readers in making their own studies of this fascinating and useful subject.

References

[1] Amann, H. (1976). Fixed point equations and nonlinear eigenvalue problems in ordered Banach spaces. *SIAM Review*, **18**, 620-709.

[2] Barta, J. (1937). Sur la vibration fondamentale d'une membrane. *Comptes Rendus de l'Academie des Sciences, Paris*, **204**, 472-73.

[3] Beals, R. (1985). Indefinite Sturm-Liouville problems and half range completeness. *Journal of Differential Equations*, **56**, 391-407.

[4] Beltramo, A. and Hess, P. (1984). On the principal eigenvalue of a periodic-parabolic operator. *Communications in Partial Differential Equations*, **9**, 919-41.

[5] Blat, J. and Brown, K. J. (1984). Bifurcation of a steady state solution in predator-prey and competition systems. *Proceedings of the Royal Society of Edinburgh*, **97A**, 21-34.

[6] Brown, K. J. and Eilbeck, J. C. (1982). Bifurcation, stability diagrams, and varying diffusion coefficients in reaction-diffusion equations. *Bulletin of Mathematical Biology*, **44**, 87-102.

[7] Brown, K. J. and Lin, S. S. (1980). On the existence of positive eigenfunctions for an eigenvalue problem with indefinite weight function. *Journal of Mathematical Analysis and Applications*, **75**, 112-20.

[8] Brown, K. J. and Lin, S. S. (1981). Bifurcation and stability results for an equation arising in population genetics. *Annali di Matematica Pura ed Applicata*, **128**, 375-87.

[9] Cantrell, R. S. and Cosner, C. (1987a). On the steady-state problem for the Volterra-Lotka competition model with diffusion. *Houston Journal of Mathematics*, **13**,337-52.

[10] Cantrell, R. S. and Cosner, C. (1987b). On the generalized spectrum for second-order elliptic systems. *Transactions of the American Mathematical Society*, **303**, 345-63.

[11] Cantrell, R. S. and Cosner, C. (preprint a). On the uniqueness and stability of positive solutions in the Lotka-Volterra competition model with diffusion.

[12] Cantrell, R. S. and Cosner, C. (preprint b). Diffusive logistic equations with indefinite weights: population models in disrupted environments.

[13] Cantrell, R. S. and Schmitt, K. (1986). On the eigenvalue problem for coupled elliptic systems. *SIAM Journal on Mathematical Analysis*, **17**, 850-62.

[14] Castro, A. and Lazer, A. C. (1982). Results on periodic solutions of parabolic equations suggested by elliptic theory. *Bolletino della Unione Matematica Italiana Serie 6*, **1-B**, 1089-104.

[15] Chow, S. N. and Hale, J. K. (1982). *Methods of bifurcation theory*. Springer Verlag, New York.

[16] Courant, R. and Hilbert, D. (1953). *Methods of mathematical physics* v. 1. Interscience, New York.

[17] Crandall, M. G. and Rabinowitz, P. H. (1971). Bifurcation from simple eigenvalues. *Journal of Functional Analysis*, **8**, 321-40.

[18] Faierman, M. and Roach, G. F. (1987). Linear elliptic eigenvalue problems involving an indefinite weight. *Journal of Mathematical Analysis and Applications*, **126**, 516-28.

[19] Fife, P. C. and Peletier, L. (1977). Nonlinear diffusion in population genetics. *Archive for Rational Mechanics and Analysis*, **64**, 93-109.

[20] de Figueiredo, D. G. (1982). Positive solutions of semilinear elliptic problems. In *Springer Lecture Notes on Mathematics*, **957**, 34-88. Springer-Verlag, New York.

[21] Fisher, R. A. (1930). *The genetical theory of natural selection. (second Edition 1958)*. Dover, New York.

[22] Fisher, R. A. (1937). The wave of advance of advantageous genes. *Annals of Eugenics*, **7**, 355-69.

[23] Fisher, R. A. (1950). Gene frequencies in a cline determined by selection and diffusion. *Biometrics*, **6**, 353-61.

[24] Fleckinger, J. and Lapidus, M. L. (1986). Eigenvalues of elliptic boundary value problems with an indefinite weight function. *Transactions of the American Mathematical Society*, **295**, 305-24.

[25] Fleckinger, J. and Mingarelli, A. B. (1984). On the eigenvalues of nondefinite elliptic operators. In *Differential equations*, ed. I. W. Knowles and R. T. Lewis, North-Holland Mathematics Studies **92**. North-Holland, New York.

[26] Fleming, W. H. (1975). A selection-migration model in population genetics. *Journal of Mathematical Biology*, **2**, 219-33.

[27] Gilbarg, D. and Trudinger, N. S. (1977). *Elliptic partial differential equations of second order*. Springer-Verlag, New York.

[28] Gossez, J.-P. and Lami Dozo, E. (1982). On an estimate for the principal eigenvalue of a linear elliptic problem. *Portugaliae Mathematica*, **41**, 347-50.

[29] Gossez, J.-P. and Lami Dozo, E. (1985). On the principal eigenvalue of a second order linear elliptic problem. *Archive for Rational Mechanics and Analysis*, **89**, 169-75.

[30] Henry, D. (1981). *Geometric theory of semilinear parabolic equations*. Lecture Notes in Mathematics, **840**. Springer-Verlag, New York.

[31] Hess, P. (1977). On uniqueness of positive solutions of nonlinear elliptic boundary value problems. *Mathematische Zeitschrift*, **154**, 17-18.

[32] Hess, P. (1982a). On bifurcation and stability of positive solutions of nonlinear elliptic eigenvalue problems. In *Dynamical systems II*. (ed. A. R. Bednarek and L. Cesari). Academic Press, New York.

[33] Hess, P. (1982b). On the principal eigenvalue of a second order linear elliptic problem with an indefinite weight function. *Mathematische Zeitschrift*, **179**, 237-39.

[34] Hess, P. (1983). On the eigenvalue problem for weakly coupled elliptic systems. *Archive for Rational Mechanics and Analysis*, **81**, 151-59.

[35] Hess, P. (1984). On positive solutions of semi-linear periodic-parabolic problems. In *Lecture Notes in Mathematics*, **1076**, 101-14. Springer-Verlag, New York.

[36] Hess, P. (1985). On the relative completeness of the generalized eigenvectors of elliptic eigenvalue problems with indefinite weight functions. *Mathematische Annalen*, **270**, 467-75.

[37] Hess, P. (1986). On the asymptotic distribution of eigenvalues of some nonselfadjoint problems. *Bulletin of the London Mathematical Society*, **18**, 181-84.

[38] Hess, P. and Kato, T. (1980). On some linear and nonlinear eigenvalue problems with an indefinite weight function. *Communications in Partial Differential Equations*, **5**, 999-1030.

[39] Hirsch, M. (1988). Stability and convergence in strongly monotone dynamical systems. *Journal für Reine und Angewandte Mathematik*, **383**, 1-53.

[40] Holland, C. J. (1978). A minimum principle for the principal eigenvalue for second order linear elliptic equations with natural boundary conditions. *Communications on Pure and Applied Mathematics*, **31**, 509-19.

[41] Lazer, A. C. (1982). Some remarks on positive solutions of parabolic differential equations. In *Dynamical systems II*, (ed A. R. Bednarek and L. Cesari). Academic Press, New York.

[42] Levin, S. (1988). Population models and community structure in heterogeneous environments. In *Mathematical biology* (ed. T. G. Hallam and S. Levin). Springer-Verlag, New York.

[43] Lotka, A. J. (1925). *Elements of physical biology.* Williams and Wilkins, Baltimore.

[44] Ludwig, D., Aronson, D. G., and Weinberger, H. F. (1979). Spatial patterning of the spruce budworm. *Journal of Mathematical Biology*, **8**, 217-58.

[45] Manes, A. and Micheletti, A.-M. (1973). Un'estensione della teoria variazionale classica degli autovalori per operatori ellitici del secondo ordine. *Bolletino della Unione Matematica Italiana Serie 4*, **7**, 285-301.

[46] Matano, H. (1984). Existence of nontrivial unstable sets for equilibriums of strongly order-preserving systems. *Journal of the Faculty of Science, University of Tokyo*, **30**, 645-73.

[47] Mingarelli, A. B. (1982). Indefinite Sturm-Liouville problems. In *Lecture Notes in Mathematics*, **964**, 519-28. Springer-Verlag, New York.

[48] Mingarelli, A. B. (1983). On the existence of nonsimple real eigenvalues for general Sturm-Liouville problems. *Proceedings of the American Mathematical Society*, **89**, 457-60.

[49] Murray, J. D. and Sperb, R. P. (1983). Minimum domains for spatial patterns in a class of reaction diffusion equations. *Journal of Mathematical Biology*, **18**, 169-84.

[50] Nussbaum, R. (1984). Positive operators and elliptic eigenvalue problems. *Mathematische Zeitschrift*, **186**, 247-64.

[51] Okubo, A. (1980). *Diffusion and ecological problems: mathematical models.* Springer-Verlag, New York.

[52] Protter, M. H. and Weinberger, H. F. (1966). On the spectrum of general second order elliptic operators. *Bulletin of the American Mathematical Society*, **72**, 251-55.

[53] Rabinowitz, P. H. (1973). Some aspects of nonlinear eigenvalue problems. *Rocky Mountain Journal of Mathematics*, **3**, 161-202.

[54] Reed, M. and Simon, B. (1978). *Methods of modern mathematical physics IV, Analysis of Operators.* Academic Press, New York.

[55] Saut, J. C. and Scheurer, B. (1978). Remarks on a nonlinear equation arising in population genetics. *Communications in Partial Differential Equations*, **23**, 907-31.

[56] Senn, S. and Hess, P. (1982). On positive solutions of a linear elliptic boundary value problem with Neumann boundary conditions. *Mathematische Annalen*, **258**, 459-70.

[57] Skellam, J. G. (1951). Random dispersal in theoretical populations. *Biometrika*, **38**, 196-218.

[58] Smith, H. (1988). Systems of ordinary differential equations which generate an order preserving flow. A survey of results. *SIAM Review*, **30**, 87-113.

[59] Smoller, J. (1983). *Shock waves and reaction-diffusion equations*. Springer-Verlag, New York.

[60] Verhulst, P. F. (1838). Notice sur la loi que la population suit dans sou accroissment. *Correspondences Mathématiques et Physiques*, **10**, 113-21.

[61] Verhulst, P. F. (1847). Deuxieme mémoire sur la loi d'accroisement de la population. *Mémoires, Académie Royale de Belgique*, **20**, 1-32.

[62] Volterra, V. (1931). *Lecon sur la théorie mathématique de la lutte pour la vie*. Gauthier-Villars, Paris.

[63] Weinberger, H. F. (1974). *Variational methods for eigenvalue approximation*. Society for Industrial and Applied Mathematics, Philadelphia.

Mathematical Aspects of Electrophoresis

Paul C. Fife
University of Utah

Xiao Geng
Brown University

I Background and Formulation

A The basic model

1. Background and introduction

The object of electrophoresis techniques is to effect a separation among individual types of charged particles existing in a solution. This is done by passing an electric current through the solution; the differing responses of the various particle types to the induced electrical field then provides a separation mechanism.

There are a great many practical ways to utilize these differences, and so there are many sorts of electrophoretic techniques. Our intention here is not to give a survey of them, but rather to discuss some mathematical issues which are basic to all or many of them. For more in the way of background, see for example Deyl (1979), Babskii *et al.* (1983), Bier *et al.* (1983), Thormann (1984, 1985), Saville and Palusinski (1986). In particular, the book (Babskii *et al.* 1983) is a valuable compendium of physical and mathematical results in this broad subject.

In this first part we shall bring out various forms which some important mathematical models of electrophoresis take. All those treated here are reductions or special cases of a single basic system of partial differential equations. In some form or another, all of these models appear in the literature cited above.

In the later parts, we shall indicate some of the basic analysis which has recently been done for these models, especially at the University of Arizona. The principal aims are to be able better to understand, from this analysis, the reason separation is possible, and to find the properties of the separation phenomena which the models predict.

This survey is not meant to be exhaustive; a great many theoretical studies of electrophoresis and related phenomena will not be mentioned. Moreover, extensive computational work has been done on these problems; that aspect will by and large be neglected. Finally, matters of historical perspective and priority will usually not be dealt with. As in all scientific fields, the important theories in this one have evolved gradually. For example, the Kohlrausch relation was known almost a century ago, but the theory surrounding it, in the state of completeness expounded here for example, was certainly not.

2. Definitions and equations

Consider a mixture of charged species in solution. We denote by u_i the concentration of species i, and by U the vector whose components are these concentrations. Each species has its characteristic mobility μ_i, charge z_i, and diffusivity d_i. In accordance with usual practice we assume the latter to be of the form $d_i = d\mu_i$, d here being a positive constant independent of i. The mobility is a measure of the degree of response which this type of particle exhibits in the presence of an electric field. These mobilities and charges can be considered as components of a mobility vector $\underline{\mu}$ and a charge vector $\mathbf{z}$ respectively. The mobilities as assumed to be ordered as

(1.1) $$\mu_1 < \mu_2 < \mu_3 < \cdots,$$

We also assume there is a variable (in space, at least) unknown electric field E present. Space, in the models discussed here, will be taken as one-dimensional. The models will also neglect any hydrodynamic or temperature effects; the only means of transport for the particles will be by their response to E and by diffusion. There may or may not be chemical reactions of interest.

In this scenario the dynamics of the functions $U(x,t)$ and $E(x,t)$ is governed by the following system, derived in Babskii *et al.* (1983), Saville and Palusinski(1986), Fife(1988) and in other places:

(1.2) $$U_t + \underline{\mu}(\mathbf{z}(EU)_x - dU_{xx}) = \mathbf{g}(U),$$

(1.3) $$\beta E_x = -\mathbf{z} \cdot U.$$

Here the vector **g** is the source term resulting from the reactions. Contiguous vector symbols, such as in the second term in (1.2), denote vectors which are the componentwise products of two or more vectors. On the other hand the dot product in (1.3) is the usual scalar product. The coefficient β in (1.3) (Poisson's equation) is a physical constant involving the permittivity of the solvent.

It is clear from (1.3) that the flux vector for this system is

(1.4) $$\mathbf{f} = \underline{\mu}(\mathbf{z}EU - dU_x).$$

Finally, the electric current through the medium is given by

(1.5) $$\mathbf{J} = -\mathbf{z} \cdot \mathbf{f}.$$

Various classes of electrophoretic phenomena are governed by particular cases of (1.2), (1.3), together with particular boundary conditions and initial conditions.

B Reactionless cases

1. Equations under the electroneutrality assumption

The total charge density is given by the scalar product on the right of (1.3). This is often assumed to vanish; in other words, one often sets $\beta = 0$. In the usual situations β is in fact very small and so formally this assumption is reasonable. The rigorous justification of this approximation has only been given, however, in the case of a prototypical travelling wave problem (III below).

We now explore the consequences of setting β and $\mathbf{g} = 0$ in (1.2), (1.3). In this case, the function E may be expressed in terms of other physical quantities; in fact take the scalar product of (1.2) with the vector $\mathbf{z}$ and use (1.3), (1.4), (1.5) to obtain

$$\mathbf{z} \cdot \mathbf{f}_x = -\mathbf{J}_x = 0$$

which means that $\mathbf{J}$ is constant in x. We assume it is also constant in time; in fact the current is controllable in the laboratory. Therefore

(1.6) $$\mathbf{z} \cdot \underline{\mu}(\mathbf{z}EU - dU_x) = -\mathbf{J} = \text{const.}$$

From this we find

(1.7) $$E = \frac{-\mathbf{J} + d\mathbf{z} \cdot \underline{\mu} U_x}{\mathbf{z} \cdot \underline{\mu} \mathbf{z} U},$$

which may now be inserted into (1.2) to obtain a system for the concentrations U alone. This, of course, is a significant simplification.

2. A further reduction in the diffusionless case

If we now set $d = 0$ in (1.2) and (1.7), we obtain

(1.8a) $$U_t - \mathbf{J}\underline{\mu} \left(\frac{\mathbf{z}U}{\mathbf{z} \cdot \underline{\mu}\mathbf{z}U} \right)_x = 0,$$

(1.8b) $$\mathbf{z} \cdot U = 0.$$

3. The case when there is a single counterion

In typical models for a separation procedure called *isotachophoresis*, one assumes that the species are all either positively or negatively charged ions; that all the ions to be separated have charges of one sign (negative, let us say); and that there is only a single type of ion of the other (positive) sign, called the counterion.

In this case we may denote by u_0 the concentration of the counterion, and by $u_i, i = 1, \cdots, n$, those of the n other species. We suppose that $z_0 = 1$ and $z_i = -1$ for $i \geq 1$. Then (1.8b) results in

(1.9) $$u_0 = \sum_{i=1}^{n} u_i,$$

and when we put this into (1.8a), we obtain, for $i = 1, \cdots, n$,

$$\text{(1.10)} \qquad \partial_t u_i + \mathbf{J}\mu_i \partial_x \left(\frac{u_i}{\sum_j (\mu_j + \mu_0)\mu_j} \right) = 0.$$

Let $v_i = (\mu_i + \mu_0)u_i$ ($i = 1, \cdots, n$) denote the components of an n-component vector V, and similarly $\alpha_i = \mathbf{J}\mu_i$. With this, , (1.10) now becomes

$$\text{(1.11)} \qquad V_t + \left(\frac{\alpha V}{\sum v_j} \right)_x = 0,$$

a system of conservation laws which turns out generally to be strictly hyperbolic, and which will be explored below in II.

4. The travelling wave problem

Here we no longer assume β and d are zero, and seek solutions of (1.2), (1.3) with $\mathbf{g} = 0$, of the form $U = U(x - ct) = U(\zeta), \zeta = x - ct$. Denoting derivatives with respect to ζ by primes, we have

$$\text{(1.12a)} \qquad -cU + \underline{\mu}(\mathbf{z}EU - dU') = \text{const.},$$

$$\text{(1.12b)} \qquad \beta E' = -\mathbf{z} \cdot U.$$

We look for solutions which approach constants $U_\pm, E_\pm$ at $\zeta = \pm\infty$, with derivatives vanishing there. The constant in (1.12a) can therefore be expressed in terms of c and these limiting values. Other than the parameters already appearing in (1.2), (1.3), there are the following numbers directly or indirectly involved in (1.12): the boundary values at $\pm\infty$ mentioned above, the current $\mathbf{J}$ (1.5), and the velocity c. They are not all independent, however; we may derive relations among them as follows. Evaluating (1.12a) at $\pm\infty$, we obtain

$$\text{(1.13)} \qquad (\mathbf{z}\underline{\mu}E_+ - c)U_+ = (\mathbf{z}\underline{\mu}E_- - c)U_-.$$

Doing the same with (1.3) and (1.5), we find

$$\text{(1.14)} \qquad \mathbf{z} \cdot U_\pm = 0,$$

$$\text{(1.15)} \qquad \mathbf{J} = -\mathbf{z} \cdot \underline{\mu}\mathbf{z}U_\pm E_\pm.$$

A laboratory setup which results in travelling waves in the form of moving separated zones starts with an initial distribution of ions such that (a) there is a single counterion as in section 3 above, present everywhere; (b) far to the left there is only one other type of ion (ion 1, called the terminating ion); and (c) far to the right there is only ion n (the leading ion) present. Recall that for $i \geq 1$ the mobilities (1.1) were arranged in increasing order, so that the terminating ion has lowest mobility of all, and the leading one has the largest. Let us consider boundary conditions corresponding to this arrangement.

We have

(1.16)
$$U_- = (u_{0-}, u_{1-}, 0, \cdots), U_+ = (u_{0+}, 0, 0, \cdots, u_{n+}),$$
$$\mathbf{z} = (1, -1, -1, \cdots).$$

Putting these values into (1.13)–(1.15), we obtain

(1.17)
$$\mu_n E_+ = -c = \mu_1 E_-, \mathbf{J} = -(\mu_1 + \mu_0) u_{1-} E_- = -(\mu_n + \mu_0) u_{n+} E_+,$$
$$u_{0-} = u_{1-}, u_{0+} = u_{n+}.$$

Hence among all eight adjustable parameters ($\mathbf{J}, c$, and the various limiting values at $\pm\infty$), there can be only two independent ones. We choose one of them to be $\mathbf{J}$, because that can be controlled. It will turn out that the concentration of the leading ion can also be controlled, so we choose $u_{n+} \equiv \sigma$ to be the other one. In terms of those two, we may solve for the rest to obtain

(1.18)
$$E_+ = -\frac{\mathbf{J}}{(\mu_n + \mu_0)\sigma}, E_- = \frac{E_+}{b_n}, c = -\mu_1 E_-,$$
$$u_{1-} = \kappa\sigma = u_{0-}, u_{0+} = \sigma,$$

where we have set

(1.19)
$$b_i \equiv \mu_1/\mu_i; \kappa = \frac{(\mu_n + \mu_0)\mu_1}{(\mu_1 + \mu_0)\mu_n} = \frac{b_0 + b_n}{b_0 + 1}.$$

The number κ is the Kohlrausch coefficient (Kohlrausch 1897); we shall encounter it later in Sec.II.D in a similar context.

We now return to (1.12a). Change to new variables W and e by

$$U = W\sigma, E = E_- e,$$

and also rescale the independent variable ζ with a suitable factor proportional to d to obtain the system

(1.20a) $$W' = -(\mathbf{z}e + \mathbf{b})W + \mathbf{q},$$
(1.20b) $$\varepsilon e' = -\mathbf{z} \cdot W.$$

Here $\mathbf{b}$ is the vector with components in (1.19); we set $q_0 = b_0 + b_n$; all other components q_i of the vector $\mathbf{q}$ are 0; and $\varepsilon = |E_-|\beta/d\sigma$ is our new small parameter. (Note that $E_- < 0$.) The boundary conditions are

(1.21)
$$W(-\infty) = (\kappa, \kappa, 0, \cdots), W(+\infty) = (1, 0, 0, \cdots, 1),$$
$$e(-\infty) = 1, e(\infty) = b_n,$$

where κ is given by (1.19). Finally, when $n > 2$, the determination of a unique solution will involve integral conditions specifying the total amounts present of each species $A_i, i = 1, 2, \cdots, n - 1$.

C Fast reactions among ampholytes

1. The scenario

Ampholyte molecules are organic molecules which can exist in several different charged states. We envisage a mixture of n sample ampholyte species $A_i, i = 1, \cdots, n$, in solution, with water as solvent. There may be other electrolytes present in the solution, as well. We suppose that each of the A_i can have, for simplicity, $z = 0$ or ± 1. For now, we designate these different forms as three different species A_i^s, $s = -, 0,$ or $+$. These same superscripts will also distinguish their mobilities, concentrations, etc.. Thus u_3^- will be denote the concentration of A_3^-.

Ampholytes can change their charge through the chemical reactions

(1.22a) $$A_i^- + \mathbf{H}^+ \rightleftharpoons A_i^0,$$

(1.22b) $$A_i^0 + \mathbf{H}^+ \rightleftharpoons A_i^+,$$

where $\mathbf{H}^+$ is the hydrogen ion. Assume there exist no other reactions among them. There is also the reversible reaction among $\mathbf{H}^+, \mathbf{OH}^-$, and $\mathbf{H_2O}$ and, possibly, reactions involving the other electrolytes. In principle, the system (1.22) may include balance equations for the concentrations of all of these species. Those concerning the sample species number $3n$.

The source term $\mathbf{g}$ can be written down in terms of U and the various rate constants for the reactions mentioned. In any case, the fact that all reactions conserve charge results in the following identity which must hold for the entire system:

(1.23) $$\mathbf{z} \cdot \mathbf{g}(U) = 0.$$

2. The grouping approximation

We now assume that the rate constants for (1.22) are so large that the reactions can be regarded as always existing in partial equilibrium; that means that at all time,

(1.24a) $$u_i^- h = K_i^0 u_i^0,$$

(1.24b) $$u_i^0 h = K_i^+ u_i^+,$$

where for simplicity we have let h equal the concentration of $\mathbf{H}^+$, and the K's are equilibrium constants.

Let $v_i = \sum_s u_i^s$ denote the total concentration of A_i in all its forms. Then (1.24) allows us to express all ampholyte concentrations in terms of the v's and h. By algebra we obtain

(1.25a) $$u_i^s = \rho_i^s v_i,$$

where

(1.25b)
$$\rho_i^s = \frac{\sigma_i^s}{\sigma}; \quad \sigma_i^- = \frac{K_i^0}{h},$$
$$\sigma_i^0 = 1, \quad \sigma_i^+ = \frac{h}{K_i^+},$$
$$\sigma = \sigma_i^- + \sigma_i^0 + \sigma_i^+.$$

We now group triples of equations in (1.2) referring to the ampholytes by summing over s for each i and using the notation v_i. The three components of $\mathbf{g}$ in each case add up to 0 (this, in fact, can be seen a priori since the total amount of each ampholyte is conserved). When we do this, the following quantities appear:

(1.26)
$$\bar{\mu}_i(h) \equiv \sum_s \rho_i^s \mu_i^s, \quad \mathbf{m}_i(h) \equiv \sum_s \rho_i^s \mu_i^s z_i^s.$$

Note that these are certain average values of μ and $z\mu$ respectively; however, they depend on h through the ρ's dependence on h. Let $\underline{\bar{\mu}}$ and $\mathbf{m}$ be the corresponding vectors.

What results is a system for the vector $V = (v_1, v_2, \cdots)$ of the form

(1.27)
$$V_t + (E\mathbf{m}(h)V)_x - d(\underline{\bar{\mu}}(h)V)_{xx} = 0.$$

(This corrects an error in Fife 1988, III.13, 21).

We now suppose that the spatial domain is finite, and since the total amount of each ampholyte is conserved in time (none of them are created or destroyed at the electrode boundaries), it is appropriate to supplement (1.27) with no-flux boundary conditions: the flux $E\mathbf{m}V - d(\underline{\bar{\mu}}V)_x$ vanishes at each endpoint.

This system is not self-contained, because the functions E and h are unknown. In general the strict determination of E and h would involve all the remaining equations in (1.2) with boundary conditions, as well as (1.3). Various approximations have been used to reduce the magnitude of the problem. We shall return to a discussion of some of them in Sec.IV.

3. Equilibrium configurations

Let us now assume that the dynamics of the system as described above has come to an equilibrium state. We will be interested in the distributions of the various ampholytes v_i in this final state. We look for steady solutions of (1.27); they clearly have the property that the flux is constant, and from the boundary conditions, we know that constant is zero. Therefore

(1.28)
$$d(\underline{\bar{\mu}}V)_x - E\mathbf{m}V = 0.$$

Note again that $E(x)$ is an unknown function, as is $h(x)$, hence $\mathbf{m}(h(x))$ and $\underline{\bar{\mu}}(h(x))$. However, we can make the important observation from (1.26) that $\underline{\bar{\mu}}$

must be a positive vector function, though the components of **m** may be of either or both signs. Generally E must be of one sign, for example, note from (1.7) and the fact that the denominator there is always positive that for d relatively small, E has sign opposite that of **J**, in the context developed there. A similar principle operates here.

Suppose now that the experimental setup is such that for some i, $m_i(h(x))$ changes sign exactly once. Then the ith component of (1.28) is such that the derivative of $\bar{\mu}_i v_i$ changes sign exactly once. This means that this function has a single extremum; if the change occurs in the right sense, then it is a maximum. Moreover , for small d that maximum will be in the form of a sharp peak. Since $\bar{\mu}_i$ will not in general have a sharp variation, we conclude that the ampholyte concentration v_i must be peaked at some location. The height of the peak is significant because the total amount of each ampholyte (integral of v_i) is conserved, while it is concentrated near a point. This peaking action is called focussing, and is a commonly used separation technique.

The focussing of v_i occurs near where $m_i = 0$. Let us now examine the definition (1.26) to write that condition in a different form. We have

$$0 = m_i(h) = \sum_s \rho_i^s \mu_i^s z_i^s = \rho_i^+ \mu_i^+ - \rho_i^- \mu_i^-.$$

In this, of course we could replace ρ by σ. The physical meaning of this relation is that the total charge on the ith ampholyte at that point is zero. For this reason, the location where the relation holds is called the *isoelectric point* $\mathbf{X}_i$ for the ith species. By the definitions (1.25), we obtain the condition

$$\mu_i^+ \frac{h}{K_i^+} = \mu_i^- \frac{K_i^0}{h}$$

or

$$h^2(\mathbf{X}_i) = \frac{\mu_i^-}{\mu_i^+} K_i^+ K_i^0.$$

The isoelectric point $\mathbf{X}_i$ is , for small d, approximately where the ith species aggregates. Note that these points depend on i, so the different ampholyte species A_i aggregate at different locations; this is what effects separation. Also note that the points are really locations where the concentration h of $\mathbf{H}^+$ takes a certain value. In other words, the pH of the solution must be a certain value.

Finally, note that there is no guarantee that the function $h(x)$ takes on the proper values, so that some of these points $\mathbf{X}_i$ may not exist; in particular one must arrange, by choice of the other electrolytes and of the amount of current, for a suitable pH gradient to be present.

II The Transient problem with no diffusion or reaction

A Elementary waves

As we discovered in Sec. I.B.3, the dynamics of the n species concentrations in this case are governed by the following first order system, where the symbol V in (1.11) is being replaced by U:

(2.1) $$U_t + \left(\frac{\alpha U}{S}\right)_x = 0.$$

Here

(2.2) $$U = (u_1, u_2, \cdots, u_n); \quad S = \sum u_i,$$

and

$$\underline{\alpha} = (\alpha_1, \alpha_2, \cdots) \quad (\alpha_1 < \alpha_2 < \cdots)$$

is a positive n-vector. (Recall that multiplication of vectors, as in the second term in (2.1), is component-wise.) Throughout this part, we restrict attention to vectors U which are nonnegative, with $S > 0$. In the case of initial data which are piecewise constant, a great deal of the material in this section is equivalent to material in (Babskii et al. 1983), where a study of the Riemann invariants associated with (2.1) is also made. The results in that book were extended in (Geng, unpublished thesis, 1989), and much of that extension is given here. For example, Geng (unpublished thesis, 1989) established the global existence result in Sec. C, as well as the asymptotic results for general initial data in D, and gave a more complete analysis of the Riemann problem, along lines similar to the present treatment.

The basic question we ask in the present section is the following : given a (left) state U^ℓ, which other (right) states U^r can be connected to it by a shock (with entropy condition), rarefaction wave, or contact discontinuity?

First, we consider the possibility of connections which are discontinuous transitions (shocks or contact discontinuities). We look for continuous one-parameter families $U(z) \equiv U^r(z)$ of such right states, with speeds $\sigma(z)$ and $U(0) = U^\ell$. It will turn out that for each such family, there is a one to one correspondence between the states $U(z)$ and $S(z)$, given by (2.2). Anticipating this, we select, as a convenient parameter,

(2.3) $$z \equiv \frac{S(z) - S(0)}{S(0)}.$$

Finally, we introduce the notation

(2.4a) $$\mathbf{v} \equiv \frac{U}{S} = (v_1, v_2, \cdots), \quad [U] \equiv U(z) - U(0),$$

and (with similar notation for jumps in $\mathbf{v}$) set

(2.4b) $$\mathbf{w}(z) \equiv [\mathbf{v}].$$

Thus from (2.3) and (2.4)

(2.5)
$$\sum v_i = 1; \quad \sum w_i = 0; \quad [U] = [Sv] = S(0)(z(\mathbf{v}(z) + \mathbf{w}(z));$$
$$S(z)/S(0) = 1 + z.$$

The shock condition for (2.1) is

(2.6)
$$\sigma[U] = [\mathbf{f}(U)],$$

where $\sigma(z)$ is the shock velocity, and

$$\mathbf{f} = \frac{\underline{\alpha}U}{S} = \underline{\alpha}\mathbf{v}.$$

Using (2.5) and setting $\rho(z) = \sigma(z)S(0)$, we write the shock condition in the form

(2.7)
$$\rho(z\mathbf{v} + \mathbf{w}) = \underline{\alpha}\mathbf{w}.$$

First, we sum over the components in (2.7) and use (2.5) to obtain

(2.8)
$$\rho z = \sum \alpha_i w_i.$$

One may solve for each component w_i in (2.7) to obtain

(2.9)
$$w_i = \frac{\rho z v_i(z)}{\alpha_i - \rho} = \frac{\rho z (v_i(0) + w_i)}{\alpha_i - \rho}.$$

Hence

(2.10)
$$w_i = \frac{\rho z v_i(0)}{\alpha_i - \rho(1+z)};$$
$$v_i(z) = v_i(0) + w_i(z) = \frac{\alpha_i - \rho}{\alpha_i - \rho(1+z)} v_i(0).$$

Putting (2.10) into (2.8) and dividing by ρz yields

$$1 = \sum_1^n \frac{\alpha_i v_i(0)}{\alpha_i - \rho(1+z)}.$$

We subtract from this the identity (see (2.5))

$$1 = \sum_1^n \frac{\alpha_i v_i(0)}{\alpha_i}$$

to obtain

(2.11)
$$0 = \rho \mathbf{H}(\rho(1+z)),$$

where

(2.12)
$$\mathbf{H}(y) \equiv \sum \frac{v_i(0)}{\alpha_i - y}.$$

For the time being, we restrict attention to the case when **v** (hence U) is strictly positive; the extension to the case of nonnegative vectors will be given afterwards. As mentioned before, the number z will serve as an arbitrary parameter. Given a value of z, we obtain from (2.11) that the possible values of $\rho = \sigma S(0)$ are given by $\rho = 0 = \sigma$, which is always a possible speed, together with the roots ρ of $\mathbf{H}(\rho(1+z)) = 0$, $\mathbf{H}$ given by (2.12).

Lemma 1 *For any $U(0) > 0$, there exist exactly $n-1$ roots y_k ($k = 1, \cdots, n-1$) of the equation $\mathbf{H}(y) = 0$, all positive. We also define $y_0 = 0$. There exist exactly n continuous families $(\sigma_k(z), U_k(z))$ ($k = 0, \cdots, n-1$) of solutions of (2.6) satisfying $U_k(0) = U(0), S(z) = S(0)(1+z)$. They are given by*

(2.13)
$$\sigma_k(z) \equiv \frac{y_k}{S(0)(1+z)},$$

(2.14)
$$U_k = U(0) + z\beta_k U(0)$$

where $\beta_0 \equiv 1$ and for $k \geq 1$, the components $\beta_{k,i}$ of the vectors β_k are given by

(2.15)
$$\beta_{k,i} = \frac{\alpha_i}{\alpha_i - y_k};$$

vector multiplication in (2.14), as usual, is componentwise.

Proof. The first statement is clear from the form of the function $\mathbf{H}(y)$, shown below. In fact, the roots y_k are captured between the α's:

(2.16)
$$\alpha_k < y_k < \alpha_{k+1}.$$

As we have seen $\sigma = 0$ is always a possible speed; we call it σ_0. The other speeds σ_k are the roots of $\mathbf{H}(\sigma S(0)(1+z)) = 0$; (2.13) follows.

To obtain (2.14), we substitute $\rho = \sigma_k S(0)$ and (2.13) into (2.10) to obtain

$$v_i(z) = \frac{\alpha_i - y_k(1+z)^{-1}}{\alpha_i - y_k} v_i(0).$$

Hence from (2.5),

$$U_i(z) = S(z)v_i(z) = S(0)(1+z)v_i(z) = \frac{\alpha_i(1+z) - y_k}{\alpha_i - y_k} U_i(0);$$

and (2.14) follows.

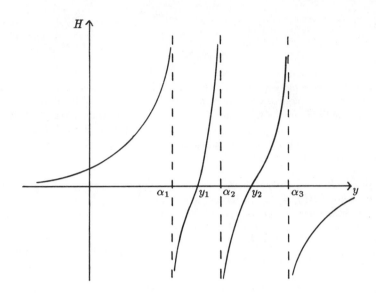

Figure 1. The function $\mathbf{H}(y)$ and its zeros.

Note from (2.14) that each family $U_k(z)$ comprises a straight line in state space passing through $U(0)$; we denote that line by ℓ_k. Clearly ℓ_0 is a ray through the origin in the direction of $U(0)$.

Let $A(U)$ denote the matrix $\mathbf{f}'(U)$. The eigenvalues λ_k of the matrix $A(U(0))$ are found by (2.13) and the relation $\lambda_k(0) = \sigma_k(0) = y_k/S(0)$. This shows the system to be strictly hyperbolic for $U > 0$. Our objective now is to find the continuations of these eigenvalues to nonzero values of z, i.e., the continuous functions $\lambda_k = \lambda_k(U_k(z))$ which are eigenvalues of $A(U_k(z))$ and which reduce to the $\lambda_k(0)$ mentioned above, when $z = 0$. Along with this, we seek the corresponding right eigenvectors $\mathbf{r}_k(z) = \mathbf{r}_k(U_k(z))$.

Lemma 2 *The formulas*

(2.17) $$\lambda_k(z) = \frac{\lambda_k(0)}{(1+z)^2};\ \lambda_k(0) = \frac{y_k}{S(0)},$$

(2.18) $$\mathbf{r}_k(z) \equiv \mathbf{r}_k(0) = \beta_k U(0)$$

hold. Also

(2.19) $$\frac{d}{dz}U_k(z) = \mathbf{r}_k(z).$$

Proof. Substitute (2.13) and (2.14) into (2.6) to obtain, for all k,

$$\frac{z}{(1+z)}\frac{y_k}{S(0)}\beta_k U(0) = \mathbf{f}(U(0) + z\beta_k U(0)) - \mathbf{f}(U(0)).$$

Differentiating with respect to z gives

$$(1+z)^{-2}\frac{y_k}{S(0)}\beta_k U(0) = A(U_k(z))\beta_k U(0).$$

Thus for all z and k,

(2.20) $$\mathbf{r}_k(U_k(z)) = \beta_k U(0);$$

(2.21) $$\lambda_k(U_k(z)) = (1+z)^{-2}\frac{y_k}{S(0)}.$$

This establishes (2.17) and (2.18). Finally (2.19) follows immediately from (2.14) and (2.18).

We now address the question of whether the solutions of (2.6) constructed in Lemma 1 are shocks which satisfy an appropriate entropy condition. We shall say that for fixed z, a solution $(\sigma_k(z), U_k(z))$ satisfies the entropy condition if

(2.22) $$\lambda_k(z) < \sigma_k(z) < \lambda_k(0).$$

We shall call such transitions from $U(0)$ to $U_k(z)$ a *k-shock*. On the other hand, a family of *k-rarefaction waves* is defined to be a function $U_k(z)$ which satisfies (2.19), $U_k(0) = U(0)$, and the inequalities opposite to (2.22). Finally, a family of contact discontinuities will be same, except that the inequalities in (2.22) are replaced by equalities.

Lemma 3 *For $k \geq 1$, the functions $U_k(z)$ in (2.14) represent states connected to $U(0)$ by entropy shocks if and only if $z > 0$, and by rarefaction waves if and only if $z < 0$. For $k = 0$, they are states connected to $U(0)$ by contact discontinuities for all $z \neq 0$.*

Proof. All the functions $U_k(z)$, for all z, satisfy the shock condition (2.6). In all cases, we know that (2.19) holds; we need only verify the appropriate inequalities or equalities. When $k = 0$, all quantities in (2.22) vanish, so we have a contact discontinuity. When $k > 0$, we have from (2.13) and (2.17) that (2.22) hold for $z > 0$ and the opposite when $z < 0$. This completes the proof.

Our next task is to extend the foregoing to vectors U which are not necessarily strictly positive. Let $R_n^+ = \{U \geq 0, S > 0\}$. The extension is accomplished by considering the given state $U \in R_n^+$ as the limits of positive states, and taking the corresponding limit of the relations found previously.

Let $U^\tau > 0$ denote a family of positive states depending continuously on the real parameter $\tau > 0$, such that the limit $\lim_{\tau \to 0} U^\tau = U^0$ exists and satisfies $S^0 > 0$. Let K be the set of indexes k such that $U_k^0 = 0$, and let $\mathbf{H}^0(y) \equiv \sum_i \frac{U_i^0}{\alpha_i - y}$, where of course the summation can equivalently be taken only over the complement of K. Let m be the number of elements in K. Then the first assertion of Lemma 1 applies to $\mathbf{H}^0$, with n replaced by $n - m$. We conclude that there are exactly $n - 1 - m$ roots of the equation $\mathbf{H}^0(y) = 0$, all positive. Letting y_k^τ be the roots of $\mathbf{H}(y; U^\tau) = 0$, we determine their limits as $\tau \to 0$.

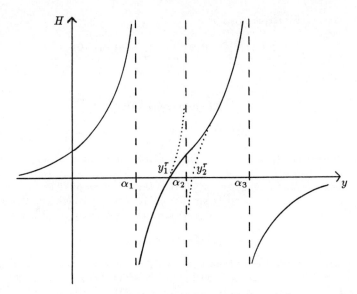

Figure 2. The functions $H^0(y)$ (solid line) and $H^0(y) + \frac{\varepsilon}{\alpha_2 - y}$ (dotted line).

Lemma 4 *For each $k \in K$,*
(a) *if $\mathbf{H}^0(\alpha_k) \geq 0$, then*

$$\lim y_k^\tau = \alpha_k; \tag{2.23a}$$

(b) *if $\mathbf{H}^0(\alpha_k) \leq 0$, then*

$$\lim y_{k-1}^\tau = \alpha_k. \tag{2.23b}$$

For all indexes k such that the above does not apply, the limit of y_k^τ is a root of $\mathbf{H}^0(y) = 0$.

Proof. This is best seen by examining the graph of $\mathbf{H}^0(y)$ when perturbed by a small term $\varepsilon(\alpha_k - y)^{-1}$ for some $k \in K$, $\varepsilon > 0$ here playing the role of U_k^τ for τ small. This perturbation (see Figure 2) introduces exactly one additional root which is close to α_k for ε small. If $\mathbf{H}^0(\alpha_k) > 0$, then the additional root is y_k^τ; in the opposite case it is y_{k-1}^τ. Finally, if equality holds, (i.e., α_k happens also to be a root of $\mathbf{H}^0(y) = 0$), then both are true; in that case we have the merging of two roots as $\tau \to 0$. But this is not possible if $k = 1$ or $k = n$.

Thus in the limit $\tau \to 0$, some of the roots y_k^τ may merge, but that is only possible in pairs, and only if one of the roots of $\mathbf{H}^0(y)$ coincides with one of the α's, $2 \leq k \leq n-1$. There are $n-2$ possible values of k for which this may happen, and for each such k, the manifold of values of U_k^0 for which this happens is exactly M_k, defined by

(2.24) $$M_k : U_k = 0; \sum \frac{U_i}{\alpha_i - \alpha_k} = 0, k = 2, \cdots, n-1.$$

Lemma 5 *For $U > 0$, the system (2.1) is strictly hyperbolic. The eigenvalues $\lambda_k(U)$ are genuinely nonlinear for $k > 0$ and linearly degenerate for $k = 0$. For $U \in R_n^+$, the same is true except on the lines M_k given by (2.24). For $U \in M_k$, $\lambda_{k-1}(U) = \lambda_k(U)$. For $U \in R_n^+$, Lemmas 1-3 still hold, with the interpretation of the y_k as limits in accordance with (2.23a) and (2.23b).*

Proof. It was already shown that the system is strictly hyperbolic for $U > 0$. The genuine nonlinearity for $k > 0$ is verified by direct calculation as follows. We designate the U-dependence of the function $\mathbf{H}$ in (2.12) explicitly, and so characterize the eigenvalues $\lambda_x(U)$ to be roots of

$$\mathbf{SH}(\lambda_k \mathbf{S}; \mathbf{U}) \equiv \sum_i \frac{U_i}{\alpha_i - \lambda_k S} = 0.$$

Differentiating with respect to U_j, we obtain

$$\sum_i \frac{U_i}{(\alpha_i - \lambda_k S)^2} \frac{\partial \lambda}{\partial U_j} + \frac{1}{\alpha_j - \lambda_k S} = 0.$$

so for some positive function $\mathbf{P}(U, \lambda_k)$, we have

$$\frac{\partial \lambda_k}{\partial U_j} = -\mathbf{P}(\alpha_i - \lambda_k S)^{-1}.$$

On the other hand from (2.18) we have $r_{k,j} = \beta_{k,j} = \alpha_j(\alpha_j - \lambda_k S)^{-1}$, so that

$$\sum_j \frac{\partial \lambda_k}{\partial U_j} r_{k,j} \neq 0.$$

This is the genuine nonlinearity condition. For $k = 0$, $\lambda_0 \equiv 0$, so the eigenvalue is linearly degenerate.

Now consider the case $U \in R_n^+$. The assertions in this case are justified by the arguments given above.

Corollary 6 *Let $U \in R_n^+$ and $u_k = 0$. If $\mathbf{H}^0(\alpha_k) \geq 0$, then*

(2.25a) $$\lambda_k(U) = \frac{\alpha_k}{S(U)}.$$

If $\mathbf{H}^0(\alpha_k) \leq 0$, then

(2.25b) $$\lambda_{k-1}(U) = \frac{\alpha_k}{S(U)}.$$

To summarize so far:

Theorem 7 *For each $U \in R_n^+ \setminus \cup M_k$, there are n linearly independent directions, specifying n straight lines $\ell_k (k = 0, \cdots, n-1)$ through U in state space, points of which are connected to U by a k-shock/rarefaction wave $(k \geq 1)$ or by a contact discontinuity $(k = 0)$.*

Some further properties of the lines ℓ_k should be brought out here.

Lemma 8 *The lines ℓ_k, for $k \geq 1$, are all perpendicular to the vector $\underline{\alpha}^{-1} = (\alpha_1^{-1}, \alpha_2^{-1}, \cdots)$. Also, for any U_0 and any k, $S(U)$ increases as one moves away from U_0 in the direction of $\mathbf{r}_k(U_0)$ (and decreases in the opposite direction). Finally, $\mathbf{r}_k(dU) = \mathbf{r}_k(U)$ for any $d > 0$.*

Proof. According to (2.18), we may write the scalar product

$$\mathbf{r}_k \cdot \underline{\alpha}^{-1} = \sum_i \beta_{k,i} u_i \alpha_i^{-1} = \sum \frac{u_i}{\alpha_i - y_k} = S(0)\mathbf{H}(y_k) = 0.$$

This proves the first assertion. The second is clear from (2.19) and the fact that by definition of z, S increases exactly when z increases.

The final assertion is evident from the following: (1) The roots y_k in Lemma 1 are independent of the magnitude of U; in fact they are given by the roots of $\sum \frac{U_i}{(\alpha_i - y)} = 0$, and this equation can be divided by $\|U\|$ with no effect. (2) The vectors β_k (2.14) are therefore also independent of $\|U\|$. (3) Therefore the directions of vectors $\mathbf{r}_k(U) = \beta_k U$ (2.19) are so independent.

B The Riemann problem and electrophoretic separation

This problem entails finding a solution of (2.1) for $t \geq 0$ with initial data

(2.26) $$U(x, 0) = \begin{cases} U^\ell, & x < 0, \\ U^r, & x > 0. \end{cases}$$

where $U^{\ell,r}$ are two given vectors, in this case in R_n^+. More than this, the object is to find such a solution in the form of a sequence of n transitions effected by elementary waves of the types studied in the previous section. Thus one wishes to identify vectors $U_k^* \in R_n^+$, $k = 0, 1, \cdots, n$, such that $U_0^* = U^\ell$, $U_n^* = U^r$, and for all k, U_k^* is connected to U_{k+1}^* by a k-shock/rarefaction wave (for $k = 1, \cdots, n-1$) or a contact discontinuity ($k = 0$). Finally, the transitions should be ordered in the $x - t$ plane, so for example if U_k^* and U_{k+1}^* are attained by shocks, then the velocities of the shocks should satisfy $\sigma_k < \sigma_{k+1}$.

According to the definition of the lines ℓ_k, we need to choose the intermediate states U^* such that U_k^* and U_{k+1}^* are on the same line ℓ_k. This could most easily be done if it were known that the n families of lines ℓ_k are the coordinate lines in some special coordinate system covering R_n^+.

Mathematical Aspects of Electrophoresis

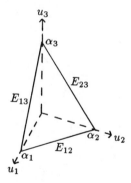

Figure 3. The part of P_1 in the positive octant.

The ℓ_0 lines are simply rays through the origin; it will therefore suffice to show that there is a family of $(n-1)$-dimensional hyperplanes in R^+ intersecting those rays transversally, each of which can be taken to have a system of coordinates given by the other ℓ's. In fact, we know (Lemma 8) that the ℓ's for $k \geq 1$ are all orthogonal to the vector α^{-1}, and in particular, every hyperplane

$$\mathbf{P}_\omega : \qquad \sum_i \frac{u_i}{\alpha_i} = \omega \ (\omega > 0)$$

has the property that any line ℓ_k with $k \geq 1$ intersecting the hyperplane lies entirely in it. Therefore it would suffice to define coordinates for each such plane by using those lines ℓ_k.

Again, it suffices to do that for only the plane $\mathbf{P}_1$. In fact, from Lemma 8, we know that the directions of the ℓ_k at a point U depend only on the direction $\frac{U}{\|U\|}$ of U, and not on its magnitude. Hence the map $T_\omega : \mathbf{P}_1 \to \mathbf{P}_\omega$ defined by $T_\omega U = \omega U$ maps ℓ_i lines onto ℓ_i lines. It therefore maps a coordinate system on $\mathbf{P}_1$ to the desired one on $\mathbf{P}_\omega$.

In the case $n = 3$, the construction of the coordinate system can be fairly easily visualized, because the part of the plane $\mathbf{P}_1$ in R_3^+ is a plane triangle, each of the three vertices lying on a coordinate axis in the original state space. (See Figure 3.)

Also for $n = 3$, there is only one singular manifold M_k (2.25), namely the one with $k = 2$. It intersects $\mathbf{P}_1$ at only one point, namely a point on the edge E_{13} where $U_2 = 0$. That one point is a singular point for the coordinate system composed of the families of ℓ_1 and ℓ_2 lines on the triangle, because the vectors $\mathbf{r}_1$ and $\mathbf{r}_2$ are linearly dependent there. Furthermore it is the only singular point. Without going into details, it turns out that the desired coordinate system has the appearance indicated in Figure 4. In the figure, the family of ℓ_1 lines intersects the segment (V_1, M_2); that of the ℓ_2 lines intersects (M_2, V_3).

It is now clear how to approach the general Riemann problem for $n = 3$. Given $U^\ell = U_0^*$, one first goes along the ℓ_0 line (ray) through that point until one reaches

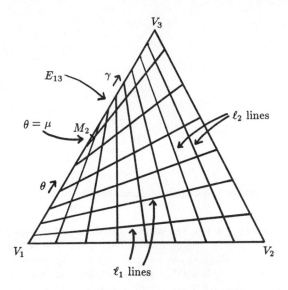

Figure 4. The coordinate system on $\mathbf{P}_1$.

the plane $\mathbf{P}_\omega$ which contains U^r. The intersection point will be U_1^*. Remaining on that plane, one then uses the special coordinate system in Figure 4 to go in two steps to the desired endpoint: first along an ℓ_1 line to a point U_2^*, and then along an ℓ_2 line to the final state. This can clearly be done in a unique manner, even when one of the given states is on M_2.

It is important to know which of these transitions are shocks, and to verify that the velocities are ordered properly. The first transition, of course, is necessarily a contact discontinuity. The other two are shocks (by Lemma 3) exactly when the parameter z introduced in (2.3) increases as one makes the transition.

To better clarify this point, it is convenient to introduce numerical coordinates in the following way. On the plane $\mathbf{P}_1$ (hence, by the natural mapping between planes, on any given $\mathbf{P}_\omega$), let us label the lines ℓ_1 according to the distance θ from the vertex V_1 to the point of intersection of that line with the edge E_{13}. Similarly, let us label the lines ℓ_2 according to the distance γ from the vertex V_1 to the point of intersection of that line with the same edge E_{13}. If the point M_2 lies at a distance μ, say, from V_1 and V_3's distance from V_1 is ρ, then the coordinate θ is restricted to the interval $[0, \mu]$, and γ to the interval $[\mu, \rho]$. The pair (θ, γ) will serve as the coordinates of points on $\mathbf{P}_1$, and hence by the ray mapping, of points on any given $\mathbf{P}_\omega$.

Consider the effect on z of traversing an ℓ_1 line. Recall from Lemma 6 that z increases in the direction of the vector $\mathbf{r}_1$. This vector of course lies in the direction of ℓ_1, but we must determine whether it points away from or toward the edge E_{13}.

It is routine to check that it points away; we therefore know that z increases as we move away from E_{13}. Traversing ℓ_1 in that direction, however, is the same as traversing it in the direction of increasing γ. We therefore have that for a fixed value of θ, z is an increasing function of γ. Likewise, it can be checked that for a fixed γ (i.e., when we remain on a fixed line ℓ_2), z is an increasing function of θ.

The following criterion follows immediately.

Lemma 9 *In the solution of the Riemann problem described above, ℓ_1-transitions for which γ increases are shocks, and the others are rarefaction waves. A similar statement holds for ℓ_2-transitions.*

It is therefore appropriate to speak of a special ordering of the states in R_3^+. The triple (θ, γ, ω) will serve as useful coordinates for points $U \in R_3^+$. The number ω specifies the triangle $\mathbf{P}_\omega$, and the other two numbers give the position on that triangle. Suppose the 2 states U^1 and U^2 have coordinates $(\theta^1, \gamma^1, \omega^1)$ and $(\theta^2, \gamma^2, \omega^2)$ in the system described above. We shall say that $U^1 \leq U^2$ if $\theta^1 \leq \theta^2$ and $\gamma^1 \leq \gamma^2$. If this relation holds between U^ℓ and U^r, then no rarefaction waves are needed in the solution of the Riemann problem. The solution involves a k-shock ($k = 1, 2$) exactly when $\gamma^1 < \gamma^2$ or $\theta^1 < \theta^2$.

Theorem 10 *For $n = 3$, if $U^\ell \leq U^r$, both in R_3^+, and the components u_2^ℓ and u_2^r not both vanishing, then the Riemann problem has a solution consisting (at most) of a contact discontinuity, a 1-shock and a 2-shock. If both shocks occur, then their velocities σ_1 and σ_2 satisfy*

$$(2.27) \qquad \sigma_1 < \sigma_2.$$

Proof. The only part of this theorem remaining to be proved is (2.27). It suffices to suppose that the U^ℓ and U^r lie on the same plane $\mathbf{P}_\omega$ (the contact discontinuity brings us immediately to that situation). So on $\mathbf{P}_\omega$, let U^* be the intermediate state (Figure 5). The shock lines ℓ_1 and ℓ_2 leading to, and issuing from, U^* may be extended (as shown by dotted lines) to points $\bar{U}^\ell$ and $\bar{U}^r$ on E_{13}.

Besides the speeds σ_1 and σ_2, we denote by $\bar{\sigma}_1$ and $\bar{\sigma}_2$ the speeds of shocks from $\bar{U}^\ell$ to U^* and from U^* to $\bar{U}^r$ respectively. Then by (2.13) and (2.17),

$$\sigma_1 = \lambda_1(U^*)(1+z) = \lambda_1(U^*)S(U^*)/S(U^\ell),$$

$$\bar{\sigma}_1 = \lambda_1(U^*)S(U^*)/S(\bar{U}^\ell),$$

and since $S(U^\ell) \geq S(\bar{U}^\ell)$, we have

$$(2.28a) \qquad \sigma_1 \leq \bar{\sigma}_1.$$

In the same way, we find

$$(2.28b) \qquad \sigma_2 \geq \bar{\sigma}_2.$$

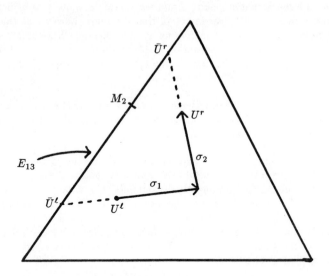

Figure 5. Shock transitions for the Riemann problem.

But we also know (again, by (2.13) and (2.17)) that

(2.29a) $$\bar{\sigma}_1 = \lambda_1(\bar{U}^\ell)S(\bar{U}^\ell)/S(U^*),$$

(2.29b) $$\bar{\sigma}_2 = \lambda_1(\bar{U}^r)S(\bar{U}^r)/S(U^*).$$

Now $\lambda_1(\bar{U}^\ell)$ is obtained from (2.25b) with $k = 2$. In fact, we calculate

$$\mathbf{H}^0(\alpha_2) = \frac{\bar{u}_1^\ell}{\alpha_1 - \alpha_2} + \frac{\bar{u}_3^\ell}{\alpha_3 - \alpha_2} < 0,$$

since this inequality defines points on E_{13} below M_2. Thus

$$\lambda_1(\bar{U}^\ell) = \alpha_2/S(\bar{U}^\ell),$$

and from (2.25a)

$$\lambda_2(\bar{U}^r) = \alpha_2/S(\bar{U}^r).$$

Hence from (2.29),

(2.30) $$\bar{\sigma}_1 = \alpha_2/S(U^*) = \bar{\sigma}_2,$$

and from (2.28),

$$\sigma_1 \leq \bar{\sigma}_1 = \bar{\sigma}_2 \leq \sigma_2.$$

At least one of the two inequalities here must be strict, unless $\sigma_1 = \bar{\sigma}_1$ and $\sigma_2 = \bar{\sigma}_2$, which is excluded by the hypothesis. This completes the proof. The following is immediate.

Corollary 11 *Consider two shocks of different type on* $\mathbf{P}_\omega$ *as in Figure 5, with common intermediate state* U^*. *Then* $\sigma_1 < \sigma_2$ *unless* U^ℓ, U^r *both lie on* E_{13}, *in which case* $\sigma_1 = \sigma_2$.

As an example, consider the case when the left state is purely $u_1 : u_2^\ell = u_3^\ell = 0$. For this case, $(\theta, \gamma) = (0, \mu)$. In constructing the solution of the Riemann problem, for the first stage (contact discontinuity with velocity 0), the only thing which appears is a discontinuity across which u_1 changes, but u_2 and u_3 remain zero. According to the diagram in Figure 4, the 2nd stage (involving an ℓ_1 line) is a transition along the edge E_{12}, which means that u_3 remains zero. This is a shock moving with positive velocity. In the final stage, which leads to the given final state, an ℓ_2 line is traversed away from the edge E_{12}, i.e., in the direction in which θ increases. It is also a shock, moving with positive speed. Therefore the Riemann problem in this case is solved by a contact discontinuity followed by a 1-shock leading to a state for which $u_3 = 0$, followed by a 2-shock leading to the given final state.

A particular case of this example provides the simplest example of separation accomplished by an electrophoretic process. Suppose the left state $U^\ell = (u_1, 0, 0)$ (in the original coordinate system) is as described, and the right state is $U^r = (0, u_2, u_3)$, where u_1, u_2, and u_3 are any positive numbers. Then the final state is on the edge E_{23}, and to attain that in the third stage, the ℓ_2 line which is used must be the one on that edge itself, i.e., the one with $\gamma = \rho$. This in turn means that the second stage is a shock connecting the left state $(u_1^*, 0, 0)$ with another pure state $(0, u_2^*, 0)$. Here the concentration u_2^* is easily determined from the fact that the second stage remains on the same plane $\mathbf{P}_\omega$; hence

$$(2.31) \qquad u_2^* = \alpha_2 u_1^*/\alpha_1.$$

Between the second and third stages (i.e., between the 1-shock and the 2-shock), the composition is pure, containing only the second ion. As time proceeds, the extent of this pure zone widens (see Figure 6). The practical effect is that although initially the second and third ions were completely mixed, the second ion becomes progressively more and more separated from the third.

This is the point at which we begin to understand the electrophoretic separation mechanism.

In this relatively simple way, one may solve a great many initial value problems with piecewise constant initial data. Other relevant examples are given in (Babskii *et al.* 1983), using a somewhat different framework. The most important example given there is that of initial data representing a finite zone in which ion 2 and 3 are uniformly mixed, pure zones representing the trailing electrolyte (ion 1) and leading electrolyte (ion 4) on either side of that finite zone. It is shown that ions 2 and 3 eventually become completely separated, occupying two contiguous zones travelling at the same positive velocity. Even though this example involves four species. it can be handled also by the methods of this section, because each shock and each interaction involves at most three species. This is the simplest illustration

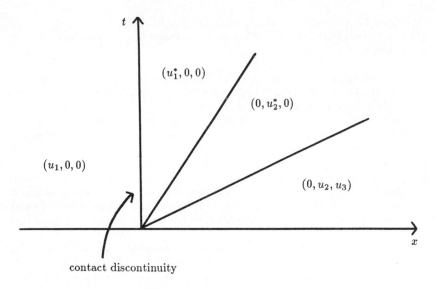

Figure 6. Shock trajectories illustrating separation.

of the isotachophoretic effect. The same effect under nonconstant initial data will be discussed in the next section.

C The initial value problem: global existence

In this section, we indicate how the method of Glimm (1965) may be applied to the existence question for solutions of (2.1) with initial data

$$U(x,0) = \Phi(x)$$

which are not necessarily piecewise constant. Still, we restrict the exposition to the case $n = 3$.

First, we give a brief description the Glimm approach to obtain approximate solutions. The spatial (x-) axis and time (t-) axis are both discretized into segments of uniform length Δx and Δt respectively. Let J_i ($i = 0, \pm 1, \pm 2, \cdots$) be the spatial segments. The ratio $\Delta t/\Delta x$ is taken to be a small enough constant ρ, to be characterized later.

At each time $t = n\Delta t$, the constructed approximation $\tilde{U}(x, n\Delta t)$ is constant on each J_i. To pass from time $n\Delta t$ to the next time $(n+1)\Delta t$, one uses an exact solution of the basic partial differential system (2.1) with initial data equal to the (piecewise constant) approximation at $t = n\Delta t$. Such a solution is found by solving a Riemann problem in a neighbourhood of each point of discontinuity, i.e., near each endpoint of a J_i. These points are spaced a distance Δx apart, and the solution consists of elementary waves with finite speed of propagation. If the ratio ρ is chosen small enough (and we so choose it), the elementary waves from neighbouring points of discontinuity will not interact while t advances a distance Δt.

This produces the desired approximation $\tilde{U}(x,t)$ for t in the time interval $n\Delta t \leq t < (n+1)\Delta t$. One then chooses a random point ξ_i (with uniform distribution) in each spatial segment J_i, and sets $\tilde{U}(x,t)$, for $t = (n+1)\Delta t$, equal to $\tilde{U}(\xi_i,(n+1)\Delta t - 0)$ for all $x \in J_i$. This gives us the piecewise data at the next time step. The process is then repeated.

This defines the continuation procedure; to start the process, one only needs data at $t = 0$ which are constant on each J_i. This approximate initial data is obtained from the exact initial data $\Phi(x)$ by the same random choice as described above.

To obtain an exact solution, an approximate solution is constructed this way for each of a sequence of values of Δx approaching zero, the ratio ρ being held constant. Under some circumstances, a subsequence will exist which converges in some sense to a limit function $U(x,t)$. Moreover this limit function will almost surely be an exact solution. The sense of the convergence is in L_1^{loc} in the upper half plane $\{t > 0\}$. See e.g. (Smoller 1983 Ch. 19 Thm.19.9).

To be assured that a convergent subsequence does exist, one needs uniform estimates on the L_∞ norm and on the total variation of the approximations. For example, the L_∞ bounds will in turn imply a uniform bound on the speeds of the elementary waves, which guarantees that a single specific value of ρ will suffice to prevent interactions. Estimates of the desired type can be obtained for general strictly hyperbolic, genuinely nonlinear, systems only if the initial data Φ has small enough total variation.

However in the case of the system being studied here, initial data without the smallness condition can be treated. That is the gist of the present section.

It is clear that a priori estimates for the total variation and uniform norm of the approximations reduce to establishing such estimates for solutions of Riemann problems. In our case, we have a good framework for understanding those solutions, namely the geometric picture described before in sec. B, together with the special coordinate system (θ, γ, ω).

First of all, it is clear that any elementary wave, therefore any solution of a Riemann problem, resides on only two planes $\mathbf{P}_\omega$, namely those of the left and right states $U^{\ell,r}$. The solution switches from one to the other by means of a contact discontinuity. Since the magnitude of a vector on $\mathbf{P}_\omega$ can be estimated in terms of ω, this fact provides a uniform bound on the magnitude $\|U\|$ of the solution in terms of the magnitude of the initial data.

It also allows us, in estimating the variation of the solution (other than the contribution from the contact discontinuity), to restrict attention to elementary waves on a single $\mathbf{P}_\omega$, so that only the coordinates θ and γ enter into the argument. This estimation is in fact accomplished by a careful and detailed look at the geometry of the ℓ_1 and ℓ_2 lines on that plane.

In this way, each approximation $\tilde{U}(x,t)$ may be estimated in respect to its L_∞

norm and total variation, in terms of the same properties of Φ. Moreover, the L_1 norm of the difference between $\tilde{U}$ at two times t_1 and t_2 may be bounded by the quantity $M|t_1 - t_2|$, M being independent of everything used in the construction, and in particular of Δx.

A global solution is then found in the usual way as the limit of a subsequence of approximate solutions.

D Asymptotic behaviour of solutions; the isotachophoretic effect

1. Step function initial data

In Theorem 10, it was brought out that a certain monotonicity assumption on the initial data of the Riemann problem with $n = 3$ results in a solution which consists only of shocks (and a contact discontinuity). This monotonicity property can be generalized to other initial data as well, and in fact the class of initial data with that property turns out to be very relevant to electrophoretic processes occurring in practice. In this section, we restrict attention to initial data of this type. For the moment we also assume it is piecewise constant.

So our assumptions on the initial data $\Phi(x) = U(x,0)$ are :

(a) Φ is piecewise constant;

(b) For some states U_+ and U_-, $\Phi(x) \equiv U_-$ for x large enough negative; $\Phi(x) \equiv U_+$ for x large enough positive:

(c) $\Phi(x_2) \geq \Phi(x_1)$ for $x_2 > x_1$.

Here, of course, the meaning of the inequality is in the sense used in Theorem 10.

First, we examine the long time behaviour of solutions with these initial data. Later, we consider analogous data which are not piecewise constant.

In the present case, the solution begins with Riemann solutions emanating from each point of discontinuity. By Theorem 10, they do not involve rarefaction waves. The shocks and contact discontinuities eventually interact. The effect of each of those interactions is another Riemann problem with initial data at the point of interaction.

It is guaranteed that the new data (after each interaction) has the same monotonicity feature, and so there still occur no rarefaction waves. By carefully evaluating the possible solution types following each kind of interaction, and tallying their cumulative effect, one is eventually able to discover that after a finite time, the solution has a very simple appearance.

First of all, it is clear from the analysis given in Section 2 that when two shocks or a shock and a contact discontinuity interact, the result has at most two of the same kinds. For example if one 2-shock overtakes a second 2-shock, all three states just before the interaction will lie on the same ℓ_2. The result after the interaction, therefore, will be a single 2-shock connecting the two outermost points on that line.

The same holds for a pair of 1-shocks. In these cases, the number of elementary waves is reduced by one. In all other cases, however, the number is preserved. For example if a 2-shock on the left overtakes a 1-shock on the right, the leftmost and rightmost states before interaction do not lie on any single ℓ_1 or ℓ_2 line. They form the initial data at the moment of interaction, so afterwards we still have a 1-shock and a 2-shock, but in opposite order. Something similar happens when a shock collides with a contact discontinuity.

Let us subdivide the time axis into the intervals between successive interactions. Generically the interactions will only be in pairs, but if three or more intersect at a single point, then we get a Riemann problem involving only the leftmost and rightmost states, and the above considerations still apply. In each interval, one may keep track of (1) the number of different states in that interval, (2) the values of θ represented among those states, and (3) the values of γ. Moreover, the changes in these three indexes when an interaction takes place may be estimated; it depends of course on the type of interaction. These results can be used to estimate the total number of interactions of each type which occur up to any given time. There turns out to be a bound on this number, depending only on the initial data.

This tells us that only a finite number of interactions are possible, again depending on that data. It means in principle that after a certain finite time t^* there will be no more, so the solution will have a particularly simple appearance.

Next, we attempt to obtain an estimate on the time t^* which is independent of the number of discontinuities in the (piecewise constant) initial function Φ. Of course $\Phi(x)$ is constant outside some finite interval (a, b). Let ω_* and ω^* be the minimum and maximum of the values of the coordinate ω represented in the range of the function $\Phi(x)$. Assume $\omega_* > 0$. It turns out that beyond the time t^* the solution consists of at most a single 1-shock, a single 2-shock, and contact discontinuities. Moreover, t^* depends only on $b - a$, ω_*, ω^*, and the states $U_\pm$.

To prove this, one first observes that all contact discontinuities have 0 velocity, so are forever confined to the interval $[a, b]$. Moreover, all shocks have positive velocity bounded away from 0 and so there must come a time (which can be estimated) after which all shocks have advanced beyond that interval. After this occurs, one considers the leftmost and rightmost 1-shock. If they are different, the difference in their speeds may be estimated from below in terms of U_+ and U_-. This is because the monotonicity requirement on the initial data allows us to control the properties of the intermediate states. Moreover, the leftmost 1-shock is faster than the rightmost one, so we can estimate the time at which they must collide. After this, there can only be a single 1-shock.

The same is true for the 2-shocks, and this finishes the proof.

So for $t > t^*$, the solution consists of some contact discontinuities to the left of b together with at most two shocks located to the right of b. For concreteness, suppose there are two. Those two shocks represent transitions on a single plane $\mathbf{P}_\omega$, namely the one on which U_+ lies. The cumulative effect of all the contact discontinuities is to project the state U_- onto that plane; let us call the image point of that projection $\hat{U}$. This latter will be the state to the immediate left of the two

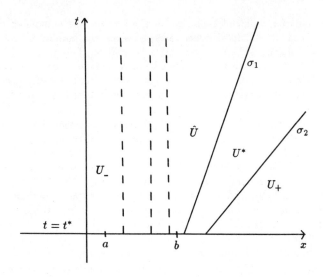

Figure 7. Shock configuration after time t^*.
Broken lines represent contact discontinuities and solid sloping lines represent shocks.

shocks. It will then be connected to U_+ by a pair of shock transitions on the plane $\mathbf{P}_\omega$, involving an intermediate state U^* between the two shocks (Figure 7).

One of the transitions is along an ℓ_1 line, the other along an ℓ_2 line. If the 1-shock is to the left of the 2-shock, then the situation is as envisaged in Corollary 11. We then conclude that $\sigma_1 < \sigma_2$ unless $\hat{U}$ and U_+ both lie on the edge E_{13}. This is compatible with the 1-shock remaining on the left. On the other hand if the 2-shock were on the left, then one could deduce from the analog of Corollary 11 that $\sigma_1 < \sigma_2$ still, which would lead to another collision later on. This has been excluded, and we conclude that the 1-shock must be on the left.

Therefore in this case, the zone belonging to the state U^* will forever be expanding at the rate $\sigma_2 - \sigma_1$. This happens exactly when either $\hat{U}$ or U_+ (or both) has nonzero component u_2. In the case of $\hat{U}$, this happens exactly when U_- has nonzero component u_2.

The conclusion is that if either U_+ or U_- has nonzero component u_2, then the two final shocks diverge as indicated. It is evident that those two shocks represent exactly the two shock transitions in the solution of the Riemann problem with initial data U_- ($x < 0$) and U_+ ($x > 0$). In other words, the asymptotic state of the solution of our original problem is the same as that of the Riemann problem, except that there are more contact discontinuities (their cumulative effect is the same as that of the single contact discontinuity produced in the Riemann problem). The intermediate states have disappeared; they affect only the positions of the final

Mathematical Aspects of Electrophoresis 165

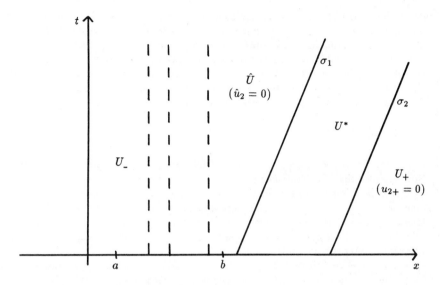

Figure 8. The isotachophoretic effect.

shocks.

2. The isotachophoretic effect

Now consider the one remaining case, namely when U_+ and U_- both have $u_2 = 0$. In this case, the second ion is initially confined entirely to $[a, b]$. Corollary 11 tells us that $\sigma_1 = \sigma_2$ in this case, so that the two shocks, if they both exist, have the same speed and move parallel to one another in the x-t plane (Figure 8). However, at this point we have not even proved that they both exist. If neither exists, then $\hat{U} = U_+$ and if only one exists, then $\hat{U}$ and U_+ are connected by a single shock along the edge E_{13}. To ensure that they do exist, we have merely to assume that the component u_2 is represented in the initial data somewhere. Then initially there will be a finite positive total amount of that ion, namely

$$\int_{-\infty}^{\infty} u_2(x,0)\, dx \; > 0.$$

The amount of each ion, if finite, is conserved, so this relation holds for all $t > 0$ as well. If there were no final pair of (parallel) shocks, there would be no u_2 remaining at that time. So both shocks do exist.

This constant velocity phenomenon is the isotachophoretic effect. The most common situation is when the states U_- and U_+ are pure (contain only the ion types 1 and n, respectively). If we begin at $t = 0$ with pure states U_- and U_+ to the far left and right respectively, and mixed states in between, the components of the mixture representing ions with mobilities between those represented in $U_\pm$, then after all interactions have taken place, the final state, apart from some motionless contact discontinuities, will consist of pure separated zones travelling with identical velocities. In the present context ($n = 3$), there will be only one such zone, consisting of the second ion. The above analysis, and in particular Corollary 11, provide the basic insight into the constant-velocity effect.

3. Other initial data

Up until now, we have considered initial data which are piecewise constant. Now suppose that $\Phi(x)$ satisfies (b) and (c) but not (a). We then approximate Φ by a step function, to which the foregoing considerations apply. Those results are independent of the fineness of the approximation. With a sequence of finer and finer approximations, a subsequence can be produced which converges to an exact solution. The latter has the same properties, for example the appearance of at most two shocks after a finite time.

Global existence theorems of this type were obtained in (LeVeque and Temple 1988) and (Serre 1987) for two-by-two systems with line characteristic fields; the asymptotic structure was not studied there, however.

4. The Kohlrausch relations

Our analysis immediately gives an explicit relationship satisfied by the concentrations in any two adjacent pure zones. Thus, suppose $U_- = (u_1, 0, 0)$ and $U_+ = (0, 0, u_3)$. After the contact discontinuities have had their effect, the pure state to the left of the travelling zone will be $(\hat{u}_1, 0, 0)$; and the state inside the travelling zone will be $(0, u_2^*, 0)$.

The fact that all states except U_- lie on the same plane $\mathbf{P}_\omega$ results in the relations

(2.32) $$u_2^* = \frac{\alpha_2}{\alpha_1} u_3 = \frac{\mu_2}{\mu_3} u_3, \quad \hat{u}_1 = \frac{\mu_1}{\mu_2} u_2^*.$$

Now recall that when the conservation laws (2.1) were first derived in the form (1.11), the change of variables $v_i = (\mu_i + \mu_0) u_i$ was made. The variables u_i in (2.32) refer to the variables in (1.11). When the relations in (2.32) are written in terms of the original variables u_i in Part I, they assume a form in which the Kohlrausch coefficients κ (1.19) appear. Specifically, let us denote the constant κ defined in (1.19) by κ_{n1}, referring to the subscripts on the mobilities other than μ_0. Similarly we may define coefficients κ_{ij} for any pair $i > j$. Then (2.32), in the original notation, becomes

(2.33) $$\hat{u}_1 = \kappa_{21} u_2^*, \quad u_2^* = \kappa_{32} u_3,$$

and hence

(2.34) $$\hat{u}_1 = \kappa_{31} u_3.$$

This last relation, together with the fourth equation in (1.18), shows that the ratio of the first and last concentration in our travelling zone, if those states are both pure, is the same as the corresponding ratio of limiting boundary values in the travelling wave problem when diffusion is present and electroneutrality not assumed. And of course from (2.33), the same Kohlrausch relation holds between any two adjacent travelling pure zones.

III Isotachophoresis with positive ε and d: the travelling wave and initial value problems

A Nature of the interface: existence questions

In Part II, the interface between two travelling zones appeared in the form of a shock. This is one result of setting d and β equal to 0. When d is small and positive, however, this interface is no longer discontinuous; it appears as a smooth but abrupt transition between the state on the left and the one on the right. This local property of a solution is often called an internal layer or the transition layer. Since these layers travel with a characteristic velocity, they are also called sharp travelling fronts. When the states to the left and right are pure states, then locally besides the counterion there are only two other ions present, and a local analysis of the layer is naturally formulated in terms of a system with $n = 2$.

In Sec.II.D it was also shown how initial value problems with initial data typical of isotachophoresis experiments result in solutions which eventually stabilize into pure zones moving with a common velocity. As mentioned, the interfaces between these zones were shocks in that context. We shall be interested in the fine structure of these shocks when they are smoothed out by diffusive action.

In view of the above, this fine structure study is best formulated in the context of the travelling wave problem with $n = 2$. We therefore consider the travelling wave equations (1.20) with that value of n. In those equations, the travelling wave coordinate ζ had been rescaled with a factor proportional to d, the result being that in (1.20a), the derivative W' is formally of the same order as the the other terms. As we shall see, the effect of this scaling is to stretch the layer so that with respect to the new travelling wave coordinate, its thickness is $O(1)$.

A complete analysis of the problem (1.20), (1.21) with $n = 2$ was given in (Fife et al. 1989), and implications of that analysis for computational methods were given in (Su et al. 1987). We shall outline the results of the former paper.

First, an existence proof was given for the problem. For this purpose the second equation (1.20b) is made more tractable by differentiating it a second time and using (1.20a). Thus

$$\varepsilon e'' = -\mathbf{z} \cdot W' = \mathbf{z} \cdot \mathbf{z} W e + \mathbf{z} \cdot \mathbf{b} W - \mathbf{z} \cdot \mathbf{q}.$$

We note that $\mathbf{z} \cdot \mathbf{z}W = \sum w_i \equiv S > 0$ and write this as

(3.1) $$\varepsilon e'' = S(e - f(W)),$$

where

$$f(W) \equiv \frac{\mathbf{z} \cdot (\mathbf{b}W - \mathbf{q})}{S}$$

is a rational function of W.

This form (3.1) of (1.20b) is more convenient for handling the apparent "singularly perturbed" form of the left side of (1.20a) where ε is small. Indeed, it turns out that the positivity of S in (3.1), the maximum principle, and the fact that our spatial domain is the entire line, guarantee that the perturbation is not singular. For example, the maximum principle applied to (3.1) tells us immediately that for any solution with W smooth (independently of ε), the function $e(\zeta)$ must satisfy $|e - f(W)| \leq O(\varepsilon)$. It can in fact be proved that any solution does have this type of smoothness, and that this order relation may be differentiated any number of times.

To proceed with the existence proof, the problem (1.20a), (3.1), (1.21) is now approximated by a certain boundary value problem on a finite interval $(-K, K)$. By virtue of a number of a priori bounds and a priori monotonicities which can be established, this latter problem is proved to have a solution by the Schauder fixed point theorem. Further detailed estimates enable us then to let $K \to \infty$ and to extract a subsequence which converges to an exact solution of the original problem. The most difficult part in this last step is establishing that the boundary conditions (1.21) are satisfied.

B Approximability for small ε

The foregoing existence result holds for all small nonnegative ε (in fact for $\varepsilon = 0$ an explicit solution can be written down). The uniqueness of the solution (modulo shifts) and its continuous dependence on ε at $\varepsilon = 0$ are proved by a totally different argument, utilizing the character of the system's critical points (the boundary values at $\pm \infty$) and a very important special property of the linearization of the right side of (1.20a) about an arbitrary state.

As a result, we may conclude that for small ε the solution may be uniformly approximated to within an error of $O(\varepsilon)$ by the explicit solution available for $\varepsilon = 0$.

It turns out, however, that for computational purposes the latter exact solution is no help; an iterative computational method for (1.20a), (3.1), devised in (Su et al. 1987) and valid for all ε, converges to a more accurate solution when $\varepsilon = 0$ than can be easily obtained by numerically evaluating the exact expression.

C The initial value problem

Consider now (1.2), (1.3) with $g = 0$, and β and d both positive. We impose arbitrary initial and boundary conditions in a class which includes conditions appropriate for isotachophoresis models. Specifically, we shall require that all functions, including the initial data, are required to approach certain limits as $x \to \pm\infty$ for all $t \geq 0$. Initial data are given only for U:

$$U(x, 0) = \Phi(x).$$

Avrin (1988, in press) has proved the existence of a unique global solution, merely requiring that Φ be locally in L_1 and that $\Phi(x) - U_\pm$ be in L_1 of a neighbourhood of $\pm\infty$; here of course $U_{\pm\infty}$ are the prescribed boundary values. Dividing (3.1) by β and integrating that equation from $-\infty$, one obtains an integro-differential equation for $U(x,t)$. This allows local existence to be proved by a contraction argument in a suitable function space. And a priori estimates allow one to extend this solution for all positive t.

This result is unusual in that the problem is nonlinear of parabolic type, yet only a very weak regularity condition (L_1) is imposed on the initial data.

It would be desirable to establish (even for $\beta = 0$) the attraction of some class of solutions of the initial value problem to the travelling wave solution. This project will be difficult; a few results related to this question were obtained in (Avrin, preprint).

IV Isoelectric focussing

The focussing phenomenon and some of the equations appropriate for modelling it were discussed in I.C. That discussion alluded to a number of difficult mathematical as well modelling problems. In this final section we shall be content with describing some (but not all) of the rigorous work which has been recently done by Su (unpublished thesis, 1989). He uses a variety of methods from nonlinear analysis. The existence results are to a great extent based on Amann's existence theory for parabolic systems (that theory is developed in many papers; the most complete results are found in the recent paper (Amann, preprint)).

1. The case of a single sample

One of the simplest models of the focussing phenomenon is the case when $n = 1$, and besides water there is no other electrolyte present. Let u_0^- be the concentration of $\mathbf{OH^-}$, and assume that ion is always in equilibrium with $\mathbf{H^+}$, in the sense that

(4.1) $$u_0^- h = K$$

for some constant K. This together with (1.27), indicates that only two concentrations, h and $v \equiv v_1$, really enter in the model. Alternatively, one could consider another buffer electrolyte to be the main source of a pH distribution.

We also assume electroneutrality, so that (1.3) holds with $\beta = 0$. We now take the scalar product of (1.2) with z. Using (1.3), differentiated with respect to t, together with (1.23),(4.1), and the grouping approximation (1.25), we obtain an algebraic expression for the unknown function E in terms of h, v, and their first derivatives in x. This reduction is similar to that leading to (1.7).

This expression, when substituted into the (grouped) balance equation (1.27) for v, results in a quasilinear parabolic equation for v with coefficients which depend on the function $h(x,t)$. When a small amount of sample is present, there may be little feedback, so the concentration of h, in other words the pH of the solution, is fixed, independent of time. So if one assumes h to be known, one obtains a simple model for the evolution of v, in the form of a single nonautonomous quasilinear equation.

The most relevant question in this case is whether the solution $v(x,t)$ approaches a peaked steady solution as $t \to \infty$. Su (unpublished thesis, 1989) has indeed demonstrated that the equation, under given initial conditions and no-flux boundary conditions, exists globally. Moreover it stabilizes to a solution of the steady equation as $t \to \infty$. A special Lyapunov functional is devised to aid in the latter result. The steady solution, of course, can easily be analyzed to see whether it exhibits the focussing effect, and it typically will if the isoelectric pH value is attained by the given function h.

2. Sample present in small concentration

This work is based on models in which there is only a single electrolyte, called the buffer, in addition to the sample ampholytes. Let us call this species **B**. Then the reaction involving **B** is

(4.2) $$\mathbf{B}^- + \mathbf{H}^+ \rightleftharpoons \mathbf{B}^0.$$

When there is no sample, the balance equations for the buffer ions, together with the electroneutrality relation and suitable boundary conditions, form a closed system. Moreover when the reversible dissociation for the buffer can be assumed in partial equilibrium, then the system reduces to a single quasilinear equation for either the concentration of $\mathbf{H}^+$ or the sum of the concentrations of the negative and neutral species. The action of the electrodes will effectively be to produce a pool of ions at the boundary of the domain, so that Dirichlet boundary conditions are appropriate.

Again, global existence and stabilization can be proved.

Now let us suppose that sample ampholyte species are present in small quantities. This situation can be expressed by representing their concentration in the form εv_i rather than v_i, where ε is a small positive parameter. Formally, the complete systems (1.2), with the sample species grouped to yield equations (1.27), appears as a perturbation of the buffer equations spoken of above.

Therefore it is natural to consider the solution obtained above for the case of no sample as the zeroth approximation to the exact solution; then seek the next approximation, which is to say a first order correction term. This correction term obeys a higher order system which is largely linear. Again, Su (unpublished thesis, 1989) has proved a priori estimates, global existence, and stabilization for this system.

Finally, the question remains as to whether there exists an exact solution for the complete system when ε is small. Such a theorem would essentially say that the solution obtained for $\varepsilon = 0$ can be continued to small positive values of ε. Su (unpublished thesis, 1989) has proved this theorem as well; the proof employs many estimates, together with the Schauder fixed point theorem. The results of Alikakos (1979) are useful.

Acknowledgment

The first author is indebted to Ken Brown and Andrew Lacey both for organizing the special year at Heriot-Watt, and for providing an extremely stimulating environment for this duration. Thanks also are due to the SERC for its financial support. The new research in this paper was supported principally by the NSF. In particular, Fife, Geng, and Su were supported under grants DMS 8503087 and DMS8703247, and in connection with Sec. III, Palusinski was supported under grant CPE 8311125. Su enjoyed the generous hospitality of Heriot-Watt University while writing part of his dissertation.

Finally, our hearty thanks go to Oleg Palusinski, who first introduecd us to the mathematics of electrophoresis.

References

[1] Alikakos, N.(1979) An application of the invariance principle to reaction - diffusion equations. *Journal of Differential Equations*, **33**, 201-25.

[2] Amann, H. Dynamic theory of quasilinear parabolic systems. III. Global existence, Preprint.

[3] Avrin, J.(1988). Global existence for a model of electrophoretic separation. *SIAM Journal of Applied Mathematics*, **19**, 520-7.

[4] Avrin, J. Viscosity solutions with singular initial data for a model of electrophoretic separation, Canadian Mathematical Bulletin, in press.

[5] Avrin, J. Some properties relating general and travelling wave solutions for a model of electrophoretic separation, preprint.

[6] Babskii, V. G., Zhukov, M. Yu. and Yudovich, V. I. (1983). *Mathematical Theory of Electrophoresis - Application to Methods of Fractionation of Biopolymers*, Naukova Dumka, Kiev.

[7] Bier, M., Palusinski, O. A., Mosher, R. A. and Saville, D. A.(1983). Electrophoresis: Mathematical modeling and computer simulations, *Science*, **219**, 1281-7.

[8] Deyl, Z.(ed.) (1979). *Electrophoresis: A survey of Techniques and Applications*, Elsevier, Amsterdam.

[9] Fife, P. C.(1988). *Dynamics of Internal Layers and Diffusive Interfaces*, CBMS-NSF Regional Conference Series in Applied Mathematics #53, SIAM, Philadephia.

[10] Fife, P. C., Palusinski, O. A. and Su, Y.(1989). Electrophoretic travelling waves, *Transactions of the American Mathematical Society*, **310**, 759-80.

[11] Geng, X.(1989). The conservation laws in the isotachophoresis model, Ph. D. Dissertation, University of Arizona, in preparation.

[12] Glimm, J.(1965). Solutions in the large for nonlinear hyperbolic systems of equations, *Communications in Pure and Applied Mathematics*, **11**, 697-715.

[13] Kohlrausch, F.(1897). Über Concentrations-Verschiebungen durch Electrolyse im Inneren von Lösungen und Lösungsgemischen, *Annalen der Physik*, Leipzig, **62**, 209.

[14] LeVeque, R. J. and Temple, B.(1985). Stability of Godunov's method for a class of 2 by 2 systems of conservation laws. *Transactions of the American Mathematical Society*. **288**, 115-23.

[15] Saville, D. A. and Palusinski, O. A.(1986). Theory of electrophoretic separations, Part 1 and Part 2, AICHE Journal, **32**, 207-23.

[16] Serre, D.(1987). Solutions a variations bornées pour certain systémes de lois de conservation, *Journal of Differential Equations*, **68**, 137-68.

[17] Smoller, J.(1983). *Shock Waves and Reaction-Diffusion Equations*. Springer Verlag, New York.

[18] Su, Y.(1989). The mathematical theory of isoelectric focussing. Ph. D. Dissertation, University of Arizona.

[19] Su, Y., Palusinski, O. A. and Fife, P. C.(1987). Isotachophoresis: analysis and computation of the structure of the ionic species interface, *Journal of Chromatograhy*, **405**, 77-85.

[20] Thormann, W.(1984). Principles of isotachophoresis and dynamics of the isotachophoretic separation of two components, *Separation Science and Technology*, **19**, 456-67.

[21] Thormann, W. and Mosher, R. A.(1986). Recent developments in isotachoporesis, in *Isotachophoresis '86*, VCH Verlagsgesellschaft, Weinheim, pp133-45.

Topological methods for the study of travelling wave solutions of reaction diffusion systems

R. A. Gardner [1]

University of Massachusetts

1 Introduction

We shall describe some topological methods which have a bearing upon the existence and also the stability of travelling wave solutions of parabolic systems of the form

(1.1) $$u_t = Du_{xx} + f(u).$$

Travelling wave solutions $u = u(x - \theta t)$ satisfy associated systems of ode's,

(1.2) $$-\theta u' = Du'' + f(u),$$

or equivalently,

(1.3) $$\begin{aligned} u' &= v \\ v' &= D^{-1}[-\theta v - f(u)] \end{aligned}$$

here "prime" is $d/d\xi$ where $\xi = x - \theta t$. Such solutions are of fundamental importance in understanding the global dynamics exhibited by (1.1) in that the stable waves supported by a particular system determine how information propagates through the one dimensional domain. The first question to resolve is to determine when a wave $u(\xi)$, θ exists with certain prescribed behaviours at $|\xi| = \infty$. A physically important case is when the wave tends to limits u_- and u_+ when ξ tends to $-\infty$ and $+\infty$. If $u_- \neq u_+$ the wave is called a *front* while if $u_- = u_+$, it is called a *pulse*. In section 2 of the paper, we shall discuss a topological invariant which is essentially the Conley index for parameterized flows, called the connection index. This index is well suited to the existence question for fronts in the physically important case where u_- and u_+ are stable rest states for (1.1). We also present some applications of these methods in the form of a singularly perturbed predator-prey system arising in mathematical ecology.

Another important issue is to determine which of the waves supported by a particular system are stable, since these are the only waves that will be physically observable. It is also of interest to be able to identify the unstable waves, since this

---

[1]This work was partially supported by NSF grant #DMS 8802468

may indicate the presence of other stable states. The local stability question for (1.1), i.e., the evolution of data consisting of a small perturbation about the wave, is determined by the spectrum of the linearized operator,

(1.4) $$Lp = Dp'' + \theta p' + df(u(\xi))p.$$

Thus we are led to the eigenvalue problem

(1.5) $$Lp = \lambda p.$$

If (1.5) admits a uniformly bounded solution $p(\xi) \not\equiv 0$ then a perturbation $p(\xi, t)$ about $u(\xi)$ formally decays or grows like $e^{\lambda t} p(\xi)$. It is well known from the theory of analytic semigroups that linearized stability/instability correctly predicts which occurs for the nonlinear equation.

In section 3 we present some new tools for counting the number of eigenvalues inside a simple closed curve K contained in a certain region of the spectral plane. These methods were introduced by Alexander, Gardner, and Jones, see (Alexander et al., 1990). In a companion paper (Gardner and Jones, 1990a) they were applied to prove the stability of travelling wave solutions to bistable predator-prey systems. More recently, these methods were extended (Gardner and Jones, 1990b) to steady state solutions of boundary value problems.

The principal constructions consist of a certain k-dimensional complex vector bundle $\mathcal{E}(K)$ over a real 2-sphere and an analytic function $D(\lambda)$, whose roots (counting order) coincide with the eigenvalues of L (counting multiplicity). The *stability index* is the first Chern number $c_1(\mathcal{E}(K))$. This invariant is equated to the eigenvalue count inside K via $D(\lambda)$. An account of the theory in (Alexander et al., 1990) is presented in section 3, together with a brief description of its application to the waves of the singularly perturbed predator-prey system obtained in Theorem 2.4 (see also (Gardner and Jones, 1990a)).

We remark that there have recently been several additions to the literature on linearized stability which rely on analytical methods based on the theory of singular perturbations (see e.g., (Nishiura and Fujii 1987; Nishiura et al., in press; Terman , in press). The geometric viewpoint developed here provides an alternative approach to this important problem.

2 Existence of solutions and the connection index

A Travelling fronts

We shall study solutions $(u(\xi), v(\xi))$ of

(2.1)
$$\begin{aligned} u' &= v \\ v' &= D^{-1}[-\theta v - f(u)] \end{aligned}$$

which satisfy the limiting conditions

$$\lim_{\xi \to \pm\infty} (u(\xi), v(\xi)) = (u_\pm, 0)$$

where $f(u_\pm) = 0$ and $u_- \neq u_+$. It will be convenient to express (2.1) in the form

(2.2) $$x' = g(x, \theta),$$

where $x = (u, v)^t \in \mathbf{R}^{2n}$ and g is the vector field in (2.1).

The case of greatest physical interest is when u_+ and u_- are both stable solutions of the reaction equations. On the level of (2.1) this forces the stable and unstable manifolds of the hyperbolic rest points

$$x_\pm = (u_\pm, 0)$$

of (2.1), which we denote by $m_\pm^s(\theta)$ and $m_\pm^u(\theta)$ respectively, to each be n-dimensional. Since a front connecting x_- to x_+ exists precisely when $m_-^u(\theta)$ intersects $m_+^s(\theta)$, we see that such an intersection is never transverse; i.e., the front can only be expected to exist at certain distinguished values of θ. Thus any invariant which can detect the connecting solution must include θ; the simplest way to achieve this is to augment (2.2) with the trivial parameter flow,

(2.3)
$$\begin{aligned} x' &= g(x, \theta) \\ \theta' &= 0. \end{aligned}$$

The centre-stable and centre-unstable manifolds of the rest points of (2.3) corresponding to x_- and x_+ are now each $n + 1$-dimensional, while the total space is $\mathbf{R}^{2n+1}$. The connection index measures whether these manifolds intersect in the augmented space. The idea for this invariant is due to Conley and Smoller (1978). This was subsequently formalized in the form presented here by Conley and Gardner (1989).

B The Connection Index

Given a flow on $\mathbf{R}^m$ and a compact neighbourhood $N \subset \mathbf{R}^m$ we denote by $S(N)$ the maximal invariant set in N under the flow. If $S = S(N)$ is isolated, i.e., $S \cap \partial N = \emptyset$, let $h(S)$ denote the Conley index of S (see (Conley 1978)).

Next, consider the parameterized flow (2.3) where $x \in \mathbf{R}^m$ and θ lies in an interval $I = \{\theta_0 \leq \theta \leq \theta_1\}$ let $X = \mathbf{R}^m \times I$ and for a given set $K \subset X$ let K_θ denote $K \cap \mathbf{R}^m \times \{\theta\}$. Noting that X_θ is invariant under (2.3), it follows that if, $\bar{N} \subset X$ then $S(\bar{N})_\theta$ (relative to (2.3)) equals $S(\bar{N}_\theta)$ (relative to (2.2)).

Now suppose that $\bar{S}, \bar{S}^0$ and $\bar{S}^1 \subset X$ are invariant under (2.3) and each family $\bar{S}_\theta, \bar{S}_\theta^0, \bar{S}_\theta^1$ of invariant sets for (2.2) are isolated and related by continuation.

Definition 2.1 *Suppose that*

(2.4)
(i) $\bar{S}^0 \cup \bar{S}^1 \subset \bar{S}$

(ii) $\bar{S}^0 \cap \bar{S}^1 = \emptyset$

(iii) $\bar{S}_\theta = \bar{S}_\theta^0 \cup \bar{S}_\theta^1$ when $\theta = \theta_0, \theta_1$.

Then $(\bar{S}, \bar{S}^0, \bar{S}^1)$ *is a* connection triple.

In most applications, we take

$$\bar{S}^0 = \{x_-\} \times I$$

$$\bar{S}^1 = \{x_+\} \times I$$

$$\bar{S} = S(N \times I)$$

where $N \subset \mathbf{R}^m$ is an isolating region for (2.2) for each $\theta \in I$ which contains x_- and x_+. In section 2. C, we shall give a simple criterion for determining connection triples.

The connection index is a tool for determining when $\bar{S}_\theta$ contains a solution running from x_- at $-\infty$ to x_+ at $+\infty$ for some $\theta \in I$. The following definition is not strictly correct; it does however provide a heuristic guide to a more precise construction given below. Thus consider subsets $\bar{N}^2 \subset \bar{N} \subset X$ such that $(\bar{N}_\theta, \bar{N}_\theta^2)$ is an index pair for $\bar{S}_\theta$ for each $\theta \in I$. Next, let $m^u(\bar{S}_\theta^0)$ denote the unstable manifold of $\bar{S}_\theta^0$. Since $\bar{S}_\theta$ does not contain connecting solutions when $\theta = \theta_0, \theta_1$, solutions in $m^u(\bar{S}_\theta^0)$ which are distinct from $\bar{S}_\theta^0$ must hit points in the exit set, $\bar{N}_\theta^2$, at $\theta = \theta_0, \theta_1$. Hence the set

$$\bar{N}^* = \bar{N}^2 \cup m^u(\bar{S}_{\theta_0}^0) \cup m^u(\bar{S}_{\theta_1}^0)$$

contains information about the global properties of $m^u(\bar{S}_\theta^0)$ as θ sweeps through the range of values in I.

Definition 2.2 *The* connection index $\bar{h}(\bar{S}, \bar{S}^0, \bar{S}^1)$ *of a connection triple is* $[\bar{N}/\bar{N}^*]$, *the homotopy type of the space obtained by collapsing* $\bar{N}^*$ *to a point.*

The following theorem is needed in order for $\bar{h}$ to be a useful, well defined index.

Theorem 2.1 (i) $\bar{h}$ *depends only on the connection triple* $(\bar{S}, \bar{S}^0, \bar{S}^1)$.
(ii) $\bar{h}$ *is invariant under continuations of the equation* (2.3) *which preserve the connection triple* $(\bar{S}, \bar{S}^0, \bar{S}^1)$, *i.e., if each of* $\bar{S}, \bar{S}^0, \bar{S}^1$ *depend on a parameter* $\lambda \in [0,1]$ *so that* $\bar{S}(\lambda)_\theta$, $\bar{S}^0(\lambda)_\theta$, *and* $\bar{S}^1(\lambda)_\theta$ *are each related by continuation and form connection triples for each* λ, *then* $\bar{h}$ *is independent of* λ.

These are two of the principal properties enjoyed by the Conley index; unfortunately, it is not possible to prove Theorem 2.1 by referring back to the usual

Conley theory because $\bar{N}, \bar{N}^*$ is not an index pair. For example, the invariant sets $\bar{S}_\theta^i$, $\theta = \theta_0, \theta_1$ lie in the boundary on $\bar{N}$, so that $S(\bar{N})$ is not interior to $\bar{N}$. This problem is addressed by perturbing the θ' equation in (2.3) to

$$\theta' = \varepsilon h(x, \theta) \ (0 < \varepsilon \ll 1)$$

where $h(x, \theta)$ is supported near $(\bar{S}_{\theta_j}^i, \theta_j)$ $j = 0, 1$. Furthermore, h is chosen to be positive (resp. negative) when (x, θ) is near $(\bar{S}_{\theta_j}^i, \theta_j)$ when $i \neq j$ (resp. $i = j$). Let $\bar{N}_j^i$ be a small neighbourhood of $(\bar{S}_{\theta_j}^i, \theta_j)$ in X chosen so that $h \neq 0$ in each $\bar{N}_j^i$, and define

$$\bar{M} = \bar{N} \cup \bigcup_{i,j=0}^{1} \bar{N}_j^i.$$

It follows that $\bar{M}$ is an isolating neighbourhood for the *perturbed* equations for all sufficiently small $\varepsilon > 0$. The exit set of $\bar{M}$ is homotopically equivalent to the set N^* (Conley and Gardner 1984). This will be illustrated in the example in section 2. E below. It follows that the usual Conley index $h(S_\varepsilon)$ of the maximal invariant set S_ε in $\bar{M}$ under the perturbed flow is well defined, independent of ε for all sufficiently small $\varepsilon > 0$, and provides the same homotopy type as $[\bar{N}/N^*]$. The properties of $\bar{h}$ in Theorem 2.1 are therefore inherited from corresponding properties of the general Conley index. It should be noted that solutions of the perturbed equations which lie in S_ε are necessarily disjoint from the support of $h(x, \theta)$; it follows that $\theta' \equiv 0$ along such solutions, and they will therefore provide solutions of the *original* equations. In this manner, we avoid an additional, and possibly quite delicate, limiting argument as $\varepsilon \to 0$.

The connection index provides a tool for measuring when $\bar{S}$ differs from $\bar{S}^0 \cup \bar{S}^1$. To this end, consider the "trivial" case where $\bar{S} = \bar{S}^0 \cup \bar{S}^1$. The connection index can easily be computed in this case by replacing the larger region $\bar{N}$ with the union $\bar{N}_0 \cup \bar{N}_1$ of neighbourhoods $\bar{N}_i$ of $\bar{S}^i$, $i = 0, 1$. The perturbed equations essentially form a product system in each $\bar{N}_i$ in which the $x-$ and $\theta-$flows decouple. From the manner in which h was chosen to change sign, the θ' equation for $\bar{N}_0$ is a repeller while it is an attractor for $\bar{N}_1$. The (Conley) indices of $S(\bar{N}_i)$ for $\varepsilon > 0$ are therefore

$$h(S(\bar{N}_0)) = \Sigma^1 \wedge h(\bar{S}^0)$$

$$h(S(\bar{N}_1)) = h(\bar{S}^1).$$

We therefore have the following theorem.

Theorem 2.2 *Suppose that* $\bar{S} = \bar{S}^0 \cup \bar{S}^1$; *then*

$$\bar{h}(\bar{S}, \bar{S}^0, \bar{S}^1) = [\Sigma^1 \wedge h(\bar{S}^0)] \vee h(\bar{S}^1).$$

Conversely, if $\bar{h}$ differs from the above index then there exists $\theta_* \in I$ such that

$$\bar{S}_{\theta_*} \supset \bar{S}_{\theta_*}^0 \cup \bar{S}_{\theta_*}^1.$$

In applications, the region $\tilde{N}$ must be constructed so that *a priori*, any solution in $\bar{S}$ is a single solution running from $\bar{S}^0$ to $\bar{S}^1$. Typically, such estimates can be obtained from a Liapunov function if one exists; in other situations, this can be obtained from monotonicity properties of certain components of the wave by building a suitable isolating neighbourhood N.

Suppose that (2.3) is augmented with a linear system $z' = Az$, where A is a hyperbolic matrix with k eigenvalues of positive real part. If $(\bar{S}, \bar{S}^0, \bar{S}^1)$ is a connection triple for (2.3), let $(\bar{S}_a, \bar{S}_a^0, \bar{S}_a^1)$ be the associated triple for the augmented system wherein all z-components are zero. It follows that

$$\bar{h}(\bar{S}_a, \bar{S}_a^0, \bar{S}_a^1) = \Sigma^k \wedge \bar{h}(\bar{S}, \bar{S}^0, \bar{S}^1).$$

This observation, together with the homotopy invariance of $\bar{h}$, is used in applications to continue higher dimensional problems to substantially simpler product systems for which $\bar{h}$ can be computed explicitly.

C An estimate for θ

The following estimate provides an *a priori* bound for the wave speed θ for solutions of (2.1). This is needed to determine a suitable interval I so that (iii) in (2.4) in the definition for connection triples will be satisfied.

Theorem 2.3 *Suppose that u_- and u_+ are stable solutions of the reaction equations,*

(2.5) $$u' = f(u),$$

and that $(u(\xi), v(\xi))$, θ is a connecting solution of (2.1) satisfying $|u(\xi)| \leq K$ for all ξ. Then there exists $L > 0$ and $\bar{\theta} > 0$ depending only on K, $f(u)$ for $|u| \leq K$, and D such that

$$|v(\xi)| \leq L \text{ and } |\theta| < \bar{\theta}$$

Proof. Set $u(x,t) = u(x - \theta t)$; then u is a solution of (1.1) and $v = u_x$. The estimates for v follow immediately from standard Schauder estimates for parabolic equations.

The bound for θ is obtained as follows. Let

$$z = Dv' = -\theta v - f(u),$$

and introduce the change of variables $v \to z$ in (2.1). The resulting system can be expressed as

$$u' = -\theta^{-1}[f(u) + z]$$

$$z' = -\theta D^{-1} z + \theta^{-1} df(u)(z + f(u)).$$

Assume first that $\theta \ll 0$. Let $U_\pm$ be an attracting neighbourhood of $u_\pm$ for (2.5). If $n_\pm(u)$ is the outward unit normal to $\partial U_\pm$ at u, there exists $\delta > 0$ depending only on $U_\pm$ such that if $|z| \leq \delta$ (where $|\cdot|$ is the max norm), then

(2.6) $$[f(u) + z] \cdot n_\pm(u) < 0$$

for all $u \in \partial U_\pm$.

Set $\delta > 0$ so that (2.6) is valid and let
$$Z(\delta) = \{z : |z| \leq \delta\}.$$

We will show that $z(\xi) \in Z(\delta)$ for all ξ provided that θ is sufficiently negative. The z-equations are
$$z' = [-\theta D^{-1} + \theta^{-1} df(u(\xi))]z + \theta^{-1} df(u(\xi))f(u(\xi)).$$

If $z \in \partial Z(\delta)$ then $|z_i| = \delta$ for some i and $|z_j| \leq \delta$ for all j. If $z_i = \delta$, then at this point
$$z'_i > |\theta| d_i^{-1} \delta - |\theta|^{-1} \{\max_{|u| \leq K} \|df(u)\| \delta + \max_{|u| \leq K} |df(u) f(u)|\}.$$

Let
$$\bar{\theta} = \{\max_i d_i (\max_{|u| \leq K} \|df(u)\| + \delta^{-1} \max_{|u| \leq K} |df(u) f(u)|)\}^{1/2};$$

then for $|\theta| > \bar{\theta}$ it follows that $\operatorname{sgn} z'_i = \operatorname{sgn} z_i$ if $|z_i| = \delta$. Since $(u(\xi), v(\xi))$ is a connecting solution it follows that $|z(\xi)| \to 0$ as $\xi \to \pm\infty$. If $z(\xi_0) \notin Z(\delta)$, the above estimate shows that $z(\xi)$ must then remain exterior to $Z(\delta)$ for all $\xi \geq \xi_0$ contradicting the decay of $z(\xi)$ at $\xi = +\infty$.

Since $z(\xi) \in Z(\delta)$, it follows from (2.6) that U_- is positively invariant for the equations
$$u' = -\theta^{-1}[f(u) + z(\xi)].$$

Since $u(\xi)$ tends to u_- at $-\infty$, $u(\xi)$ lies in U_- for all sufficiently negative ξ, and so, $u(\xi)$ remains in U_- for all ξ, contradicting the fact that $u(\xi)$ tends to u_+ at $\xi = +\infty$.

A similar estimate for large positive θ is obtained by replacing U_- with an attracting neighbourhood U_+ of u_+ in the above. In this case $Z(\delta)$ is positively rather than negatively invariant for $z(\xi)$ provided that $\theta > \bar{\theta}$; this contradicts the limiting behaviour of $u(\xi)$ as $\xi \to -\infty$.

Remark. The bound $\bar{\theta}$ for θ remains finite if some of the diffusion coefficients d_i are allowed to tend to zero, as is the case in singular perturbation problems.

D The scalar equation

We will consider the bistable diffusion equation

(2.7)
$$\begin{aligned} u' &= v \\ v' &= -\theta v - f(u) \end{aligned}$$

where f is a qualitative cubic with roots at $u = 0, \alpha, 1$ with $0 < \alpha < 1$ and $f'(0)$, $f'(1) < 0$. The phase plane for (2.7) is depicted in Figure 1 for several typical values

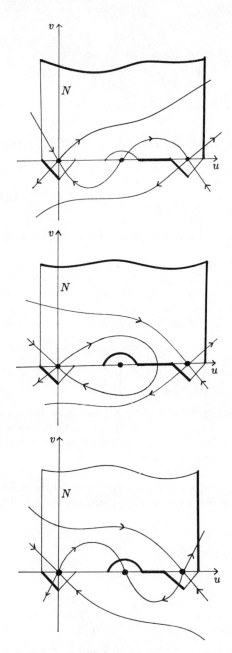

Figure 1. Phase planes associated with equation (2.7) for different values of θ.

of θ. Let N be the region indicated in the figure. Note that neighbourhoods of the saddles $(0,0)$ and $(1,0)$ have been included in N while a small neighbourhood of $(\alpha, 0)$ has been excised. It is easily seen from the hyperbolicity of the outer two rest points that any nonconstant solution in N must remain in the region $v = u' > 0$ and so, any such solution necessarily connects $(0,0)$ at $-\infty$ to $(1,0)$ at $+\infty$.

It is easily checked that if the excised neighbourhood of $(\alpha, 0)$ is small enough, and if the "top" of N is sufficiently high, then N is an isolating neighbourhood for (2.7) for all θ.

Let $\bar{\theta} > 0$ be chosen as in Theorem 2.3, let $I = \{|\theta| \leq \bar{\theta}\}$, and define $\bar{N} = N \times I$. Finally, put

$$\bar{S} = S(\bar{N})$$

$$\bar{S}^i = \{(i, 0)\} \times I, \quad i = 0, 1;$$

then $(\bar{S}, \bar{S}^0, \bar{S}^1)$ is a connection triple. The set $\bar{N}^2$ is indicated by the heavy lines in the Figure 1; for all θ $\bar{N}_\theta^2$ consists of three distinct components. Since the unstable manifolds $m^u(S^0(\theta_0))$ and $m^u(S^0(\theta_1))$ connect different pairs of components of $\bar{N}^2$ it follows that the set $\bar{N}^*$ is contractible to a point. Thus the homotopy type of $\bar{N}/\bar{N}^*$ is $[3cell/point] = [point] = \bar{0}$. Since the "trivial" index in this example is

$$\Sigma^1 \wedge \Sigma^1 \vee \Sigma^1 = \Sigma^2 \vee \Sigma^1 \neq \bar{0}$$

we see from Theorem 2.1 that $\bar{S} \supset \bar{S}^0 \cup \bar{S}^1$, and so, there exists $\theta_* \in I$ for which $\bar{S}_{\theta_*}$ contains a nonconstant solution. Since $u' > 0$ for solutions in $S(N)$, this solution must be a connecting orbit.

E Predator-prey systems

We next consider the pair of equations

(2.8)
$$u_{1t} = d_1 u_{1xx} + u_1 f_1(u)$$
$$u_{2t} = d_2 u_{2xx} + u_2 f_2(u),$$

where $u = (u_1, u_2)$. We assume that f_1 and f_2 satisfy

(H1) The null sets $f_1 = 0$ and $f_2 = 0$ are qualitatively as depicted in Figure 2, and $\partial f_1/\partial u_2 < 0$, $\partial f_2/\partial u_1 > 0$.

Thus u_2 is the density of a predator and u_1 is the density of its prey.

The reaction equations admit four rest points; the rest point O at the origin and C interior to the positive quadrant are stable. The existence of travelling waves connecting O to C was proved by Gardner (1984) under the following additional hypotheses on the vector field and the diffusion rates:

(H2) There exists a nested family of contracting rectangles Σ_τ, $0 < \tau \leq 1$ about the rest point at C, $\Sigma_\tau \subset \Sigma_\sigma$ for $\tau < \sigma$.

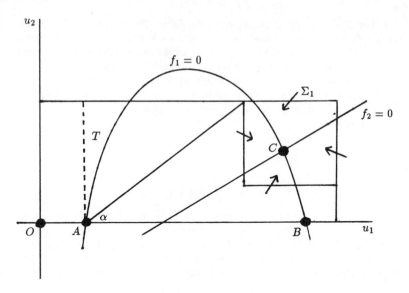

Figure 2. The null sets of f_1 and f_2.

(H3) Let θ_1 be the wave speed for the connection from O to B along which $u_2 \equiv 0$, and let $c_1 > 0$ be the smallest wave speed for connections from B to C; we assume that $\theta_1 < c_1$.

(H4) Let T be the shaded triangular region in Figure 2 to the left of $f_1 = 0$ and let
$$\varphi = \min_{u \in T}\{\arg(-f(u))\} \in (0, \frac{\pi}{2}]$$
Let α be the angle indicated in the figure. We assume that
$$d_1 > d_2 \frac{\tan \alpha}{\tan \varphi}.$$

We remark that (H2) implies that C is stable relative to (2.8); it is used to damp out oscillations that may develop when the wave is near C. (H3) ensures that the $O \to C$ connection is not split into a stacked family of waves, $O \to B \to C$. This hypothesis will be satisfied for all d_1 and d_2 if $\theta_1 < 0$, which will be the case if $\int_0^b u_1 f_1(u_1, 0) du_1 < 0$, where $B = (b, 0)$. The last condition is used to force the $u(\xi)$-components of the wave to be monotone increasing whenever $u(\xi) \notin \Sigma_1$; while this condition is somewhat artificial, some such condition is needed to isolate the $O \to C$ connection away from periodic wave trains. Note that if the left branch of $f_1 = 0$ coincides with a vertical line through A, then $\varphi = \pi/2$ and (H4) is satisfied for all d_1 and d_2; (see Figure 2)

Under hypothesis (H1)–(H4) it is proved (Gardner, 1984) that (2.8) admits waves running from O to C. The main issue is the construction of a suitable isolating region N for the associated (four dimensional) travelling wave equations.

The set N is the union
$$N = N_0 \cup N_* \cup N_c$$
of three regions in $\mathbf{R}^4$. The set N_0 is a small neighbourhood of the origin in $\mathbf{R}^4$, while N_c is the somewhat larger neighbourhood
$$N_c = \{(u,v) : u \in \Sigma_1, |v_i| \leq L, i = 1, 2\}$$
of C. The transitional region N_* is
$$N_* = \{(u,v) : u \in R, 0 \leq v_i \leq L\},$$
where R is a large rectangle in the nonnegative u-plane with vertex at the origin, and L is a large positive constant. Note that the u-components in $S(N)$ are monotone whenever $x \in N - (N_0 \cup N_c)$.

The connection index $\bar{h}$ of the associated triple is computed by suitably deforming the nonlinearities; in fact, the problem can be continued to a product system consisting of the scalar, bistable travelling wave system (2.7) and a linear, two-dimensional saddle. Thus the connection index $\bar{h}$ for the full system is $\Sigma^1 \wedge \bar{0} = \bar{0}$, which differs from the "trivial" index for this problem, $\Sigma^3 \vee \Sigma^2$. We refer to (Gardner 1989) for additional details.

F A singular perturbation problem

In this section we will prove an existence theorem for travelling wave solutions of the system (2.8) when $d_1 = \varepsilon^2$ is a small parameter and $d_2 = 1$. In this regime hypotheses (H2)–(H4) may be removed. In this scaling it is appropriate to look for waves with velocity order of ε; this suggests the introduction of a variable $\zeta = x - \varepsilon \theta t$, which yields the system

(2.9)
$$\begin{aligned} \varepsilon \dot{U}_1 &= V_1 \\ \varepsilon \dot{V}_1 &= -\theta V_1 - U_1 f_1(U) \\ \dot{U}_2 &= V_2 \\ \dot{V}_2 &= -\varepsilon \theta V_2 - U_2 f_2(U). \end{aligned}$$

The solutions to be constructed exhibit two distinct time scales; (2.9) can be used to describe the slow regions. The behaviour in transition layers is described by the stretched variables
$$\xi = \zeta/\varepsilon$$
$$u_i(\xi) = U_i(\varepsilon \xi), v_i(\xi) = V_i(\varepsilon \xi), i = 1, 2,$$

which satisfy

$$u_1' = v_1$$

$$v_1' = -\theta v_1 - u_1 f_1(u)$$

(2.10)

$$u_2' = \varepsilon v_2$$

$$v_2' = \varepsilon[-\varepsilon\theta v_2 - u_2 f_2(u)].$$

With a slight abuse of notation, we shall denote the rest points of (2.9), (2.10) for which $u = O, C$, $v = 0$ by O and C. We shall also use the notation $X = (U_1, V_1, U_2, V_2)^t$ and $x = (u_1, v_1, u_2, v_2)^t$.

Theorem 2.4 *Suppose that f_1, f_2 satisfy* (H1) *and that $d_1 = \varepsilon^2$, $d_2 = 1$. Then for all sufficiently small $\varepsilon > 0$ there exists a travelling wave solution $x(\xi, \varepsilon)$, $\theta(\varepsilon)$ of (2.10) which connects O at $\xi = -\infty$ to C at $\xi = +\infty$. The wave velocity is $\varepsilon\theta(\varepsilon)$, where $\theta(\varepsilon)$ tends to a finite limit θ^* as $\varepsilon \to 0$.*

We shall sketch the proof. First, select $\bar{f}_1$, $\bar{f}_2$ such that (H1)–(H4) are valid for all $\varepsilon \in (0, 1]$; it was indicated in the previous section how this can be achieved. In (Gardner 1989) a large isolating region N containing the rest points O and C was constructed for (2.10) which is isolating for all $\varepsilon \in (0, 1]$ for such $\bar{f}_1$, $\bar{f}_2$. Furthermore, by the remark following Theorem 2.3, there is a uniform bound $\bar{\theta}$ on the wave velocity $|\varepsilon\theta|$ which is independent of $\varepsilon \in (0, 1]$. It follows that $\bar{N} = N \times \{|\theta| \leq \bar{\theta}\}$ determines a connection triple $(\bar{S}, \bar{S}^0, \bar{S}^1)$ for all such ε. The connection index of this triple is therefore $\bar{h} = \bar{0}$ for all $\varepsilon \in (0, 1]$. Hence the theorem is valid for $\bar{f}_1$, $\bar{f}_2$ for all $\varepsilon \in (0, 1]$.

Next, we prove an *a priori* estimate which is satisfied by all solutions of (2.10) which lie in $\bar{N}$. Crude estimates are easily obtained from the manner in which $\bar{N}$ is constructed; e.g., in the fast scaling (2.10) $|u|$, $|v|$, and $|\theta|$ are uniformly bounded for $\varepsilon \in (0, 1]$, and the v components are positive whenever $u \notin \Sigma_1$. From the latter property it follows that $S(\bar{N})$ consists of rest points and connections from O to C. These estimates can be improved considerably by constructing a singular limit at $\varepsilon = 0$. The slow singular limit lies in $V_1 = U_1 = 0$ when $\zeta < 0$, and in $V_1 = 0$ and the right hand branch of $f_1 = 0$ when $\zeta > 0$ (see Figure 4): we denote this branch by $u_1 = p(u_2)$; setting

$$g_-(U_2) = U_2 f_2(0, U_2)$$

$$g_+(U_2) = U_2 f_2(p(U_2), U_2),$$

we obtain the slow reduced equations

(2.11)

$$\dot{U}_{2R} = V_{2R}$$

$$\dot{V}_{2R} = -g_\pm(U_{2R}),$$

Reaction diffusion systems

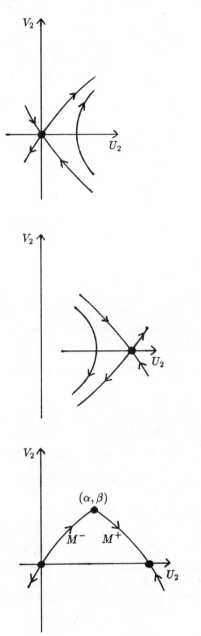

Figure 3. Phase planes for the slow reduced equations.

The phase planes for (2.11) are depicted in Figure 3. Note that the unstable manifold M^- of the origin in the $(-)$ plane and the stable manifold M^+ of the rest point on the positive U_2 axis in the $(+)$ plane intersect at a unique point $(U_2, V_2) = (\alpha, \beta)$. We parameterize solutions $(U_{2R}^\pm(\zeta), V_{2R}^\pm(\zeta))$ of (2.11) coinciding with $M^\pm$ so that they both equal (α, β) at $\zeta = 0$. The *slow singular limit* is

$$X_R(\zeta) = \begin{cases} (0, 0, U_{2R}^-(\zeta), V_{2R}^-(\zeta)), & \zeta < 0 \\ (p(U_{2R}^+(\zeta)), 0, U_{2R}^+(\zeta), V_{2R}^+(\zeta)), & \zeta > 0 \end{cases}$$

In order to describe the behaviours in the layer, set $\varepsilon = 0$ in the stretched system (2.10) to obtain

(2.12)
$$\begin{aligned} u'_{1R} &= v_{1R} \\ v'_{1R} &= -\theta v_{1R} - u_1 f_1(u_1, u_2); \\ u'_{2R} &= 0 \\ v'_{2R} &= 0; \end{aligned}$$

here u_2 is a parameter since $u'_2 = 0$. For each fixed $u_2 \in [0, \bar{u}_2)$, $u_1 f_1(u_1, u_2)$ is a qualitative cubic of the type appearing in (2.7) and so, for each such u_2 there exists a wave velocity $\Theta(u_2)$ for which there is a connecting solution of (2.12) from $(0,0)$ to $(p(u_2), 0)$. Let $u_{1R}(\xi), v_{1R}(\xi)$ be the connecting solution with $u_2 = \alpha$; the *fast singular limit* is

$$x_R(\xi) = (u_{1R}(\xi), v_{1R}(\xi), \alpha, \beta)$$

$$\theta^* = \Theta(\alpha)$$

The *singular limit* is the union of all points on $X_R(\zeta)$ and $x_R(\xi)$; its projection on the u-plane is indicated by the dotted curve in Figure 4. Given $\delta > 0$, define

$$\begin{aligned} n_\pm(\delta) &= \{x \in \mathbf{R}^4 : \text{dist}\,(x, X_R(\zeta)) \leq \delta \text{ for some } \zeta\} \\ n_F(\delta) &= \{x \in \mathbf{R}^4 : \text{dist}\,(x, x_R(\xi)) \leq \delta \text{ for some } \xi\} \\ I(\delta) &= \{\varepsilon\theta \in \mathbf{R} : |\theta - \theta^*| \leq \delta\}; \end{aligned}$$

we then set

$$N(\delta) = [n_-(\delta) \cup n_F(\delta) \cup n_+(\delta)] \times I(\delta).$$

This is a neighbourhood of the singular limit; it is easily checked that $N(\delta) \subset \bar{N}$ for all sufficiently small $\delta > 0$. It then follows that $S(N(\delta)) \subset S(\bar{N})$.

The principal estimates consist of the showing firstly that

(2.13) $$S(\bar{N}) \subset S(N(\delta)),$$

and secondly that $N(\delta)$ is an isolating region for (2.10) for all sufficiently small $\varepsilon \in (0, \varepsilon_0]$. Assuming (2.13) for the moment, it then follows that $N(\delta)$ determines

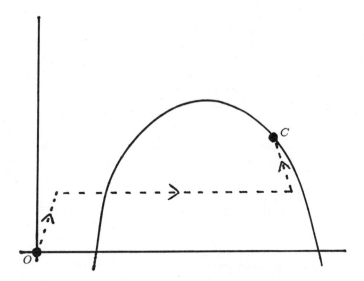

Figure 4. Projection of the fast singular limit.

the same connection triple as $\bar{N}$, so that its connection index $\bar{h}$ is the nontrivial index $\bar{0}$. Next, we note that the construction of the singular limit, and hence of $N(\delta)$, depends only on the vector field satisfying (H1), i.e., that the null sets of f_1 and f_2 are qualitatively as depicted in Figure 2. Given any such vector field $F(u) = (u_1 f_1(u), u_2 f_2(u))$, we can construct a homotopy $F(u, \lambda)$ connecting $F(u)$ at $\lambda = 0$ to $\bar{F}(u) = (u_1 \bar{f}_1(u), u_2 \bar{f}_2(u))$ at $\lambda = 1$ through a family of fields $F(u, \lambda)$ which satisfy (H1) for $\lambda \in (0, 1]$. Assuming the isolating character of the neighbourhood $N(\delta, \lambda)$ of the singular limit for the λth field, the nontriviality of $\bar{h}$ is therefore established by homotopy invariance. Thus $S(N(\delta))$ contains a nonconstant solution for some $\theta(\varepsilon) \in I(\delta)$ and for all sufficiently small $\varepsilon \in (0, \varepsilon_0]$. It then easily follows from monotonicity properties of the components of solutions in each of the tubes $n_-(\delta)$, $n_F(\delta)$, and $n_+(\delta)$ that this solution must be an $O \to C$ connection.

It remains to establish (2.13) and the isolating character of $N(\delta)$. Suppose then that $x(\xi, \varepsilon)$, $\theta(\varepsilon)$ is a solution of (2.10) in $S(\bar{N})$ (resp. $S(N(\delta))$), and parameterize $x(\xi, \varepsilon)$ so that

(2.14) $$u_1(0, \varepsilon) = h$$

for some fixed $h \in (0, p(\alpha))$. Here, h is chosen so that $u(0, \varepsilon) \notin \Sigma_1$ (resp. $x(0, \varepsilon) \notin n_-(\delta) \cup n_+(\delta)$). This condition forces $v_1(0, \varepsilon) = u_1'(0, \varepsilon)$ to be positive. In either case, we will show that this implies that for small ε, $x(\xi, \varepsilon)$ approximates the solution of (2.12) which connects the left slow manifold $u_1 = 0$ at $\xi = -\infty$ to the right slow manifold, $u_1 = p(u_2)$ at $\xi = +\infty$.

To this end let $\hat{x}$, $\hat{\theta}$ be a limit point of $x(\hat{\xi}, \varepsilon)$, $\theta(\varepsilon)$ as $\varepsilon \to 0$, where $\hat{\xi}$ is fixed. We first claim that $\hat{\theta} = \Theta(\hat{u}_2)$ where $\Theta(u_2)$ is the wave speed for (2.12). By condition (2.14), $\hat{x}$ is uniformly bounded away from the rest points of (2.12).

If $\hat{\theta} \neq \Theta(\hat{u}_2)$, the (u_1, v_1) components of the solution of (2.12) passing through $\hat{x}$ at $\xi = \hat{\xi}$ would fail to be a connecting solution. It can then be seen from Figure 1 that this solution would necessarily enter a region of phase space whose projection on the (u_1, v_1) plane is disjoint from that of both $\bar{N}$ or $N(\delta)$ in at least one time direction. Since this occurs in finite time it would then follow from Gronwall's inequality that $x(\xi, \varepsilon)$ enters such a region in finite time for sufficiently small ε and so, exits the neighbourhood $\bar{N}$ (resp. $N(\delta)$), providing a contradiction.

Next, given $\delta_1 > 0$ we claim that there exists $\bar{\xi} > 0$ and $\varepsilon_0 > 0$ such that

(2.15)
$$|u_1(\xi, \varepsilon)| + |v_1(\xi, \varepsilon)| < \delta_1 \qquad \text{for } \xi \leq -\bar{\xi}$$
$$|p(u_2(\xi, \varepsilon)) - u_1(\xi, \varepsilon)| + |v_1(\xi, \varepsilon)| < \delta_1 \quad \text{for } \xi \geq \bar{\xi};$$

i.e., $x(\xi, \varepsilon)$ remains uniformly near the left (resp. right) slow manifold for $\xi \leq -\bar{\xi}$ (resp. $\xi \geq \bar{\xi}$). By the parameterization set in (2.14) and the argument in the previous paragraph, $x(\xi, \varepsilon)$, $\theta(\varepsilon)$ must approximate a connecting solution of (2.12) uniformly on compact ξ-intervals. It follows from this and Gronwall's inequality that (2.15) holds at $\xi = \pm\bar{\xi}$ for some fixed $\bar{\xi} > 0$ determined by the connecting solution of (2.12). In order to establish (2.15) for all $|\xi| \geq \bar{\xi}$ we observe that both $\bar{N}$ and $N(\delta)$ have the property that, if x is a point in either neighbourhood which is sufficiently distant from the slow manifolds, then $v_1 = u_1' > 0$. In the case of $\bar{N}$ this occurs whenever $x \in \bar{N} - (N_0 \cup N_c)$, while in the case of $N(\delta)$ this occurs whenever $x \in N(\delta) - (n_-(\delta) \cup n_+(\delta))$. This property implies that transition layers can only occur in solutions in $S(\bar{N})$ and $S(N(\delta))$ which jump from left to right and not from right to left, since in the latter case, the positivity of v_1 for u in the transitional region would be violated. Hence $x(\xi, \varepsilon)$ can have at most one transition layer, which goes from left to right.

Having established (2.15) at $|\xi| = \bar{\xi}$ we claim that the uniqueness of the transition layer then implies (2.15) for all $|\xi| \geq \bar{\xi}$ and all sufficiently small ε. If this were not the case, there would exist a sequence $\varepsilon_n \to 0$ and $|\xi_n| \geq \bar{\xi}$ such that the solution $x(\xi_n, \varepsilon_n)$, $\theta(\varepsilon_n)$ lies at a distance of at least δ_1 from the slow manifolds. If $(\bar{x}, \bar{\theta})$ is a limit point of $x(\xi_n, \varepsilon_n)$, $\theta(\varepsilon_n)$ we have as before that the solution of (2.12) through $\bar{x}$, $\bar{\theta}$ is an orbit connecting the two slow manifolds. By construction this gives rise to a transition layer in $x(\xi, \varepsilon_n)$ which is distinct from that occurring in the interval $|\xi| \leq \bar{\xi}$, contradicting the uniqueness of the transition layer.

Now suppose that $\hat{x}$, $\hat{\theta}$ is a limit point of $x(0, \varepsilon)$, $\theta(\varepsilon)$ as $\varepsilon \to 0$, where $x(\xi, \varepsilon)$, $\theta(\varepsilon)$ is a solution in $S(\bar{N})$ (resp. $S(N(\delta))$. We have already proved that $\hat{\theta} = \Theta(\hat{u}_2)$. We now claim that $(\hat{u}_2, \hat{v}_2) = (\alpha, \beta)$ where (α, β) is the point where M_- meets M_+ (see Figure 3). If this were not the case at least one of the solution curves of (2.11) with g_+, $\zeta > 0$ and (2.11) with g_-, $\zeta < 0$ would fail to coincide with M_+ or M_-. The solution through $(\hat{u}_2, \hat{v}_2)$ would therefore be carried into a region in the u_2, v_2

plane disjoint from the projections of $\bar{N}$ and of $N(\delta)$ on this plane. Switching to the slow time scale, it would then follow from (2.15) and Gronwall's inequality that the (U_2, V_2)-components of $X(\zeta, \varepsilon)$ uniformly approximate those of the solutions of (2.11) on finite ζ-intervals disjoint from $\zeta = 0$, leading to a contradiction. Thus $(\hat{u}_2, \hat{v}_2) = (\alpha, \beta)$, so that $\hat{\theta} = \theta^* = \Theta(\alpha)$.

It can now be shown that every solution $x(\xi, \varepsilon)$, $\theta(\varepsilon)$ in $S(\bar{N})$ (resp. $S(N(\delta))$) is interior to $N(\delta)$ for all sufficiently small ε. In particular, it follows from an arguments similar to the above that any limit point $(\hat{x}, \hat{\theta})$ of $(x(\xi, \varepsilon), \theta(\varepsilon))$ as $\varepsilon \to 0$ must lie on the singular limit. Two cases must be considered separately, namely, (i) $\hat{x}$ is bounded away from the slow manifolds, and (ii) $\hat{x}$ lies near the left or right slow manifolds. In case (i) we see from an argument similar to the one used to prove that $\hat{\theta} = \theta^*$ that $\hat{x}$ must lie on the fast singular limit, $x_R(\xi)$. In the latter case we use an argument similar to the one used to prove that $\theta^* = \Theta(\alpha)$ to show that $\hat{x}$ lies on the slow singular limit, $X_R(\zeta)$. For brevity's sake, we shall omit the details.

This completes the proof that $S(\bar{N}) = S(N(\delta))$ for small ε, and that $N(\delta)$ is isolating at each step of the homotopy. This completes the proof of Theorem 2.4.

The method of proof also gives an estimate for the location of the connecting solution in phase space.

Corollary 2.5 *Let $x(\xi, \lambda)$, $\theta(\varepsilon)$ be the connecting solution of (2.4) parameterized as in (2.14). Then for all sufficiently small $\varepsilon > 0$ there exists $a > 0$ such that*

$$|x(\xi, \varepsilon) - x_R(\xi)| < \delta \quad \text{for } |\xi| \leq a$$

$$|X(\zeta, \varepsilon) - X_R(\zeta)| < \delta \quad \text{for } |\zeta| \geq a.$$

3 The stability index

In this section we shall sketch the construction of another topological invariant which measures the number of eigenvalues λ of the linearization L of (1.1) about a travelling wave solution. We shall also briefly indicate how this invariant can be used to prove the stability of the singularly perturbed waves whose existence was proved in Theorem 2.4.

A Linearized stability

The equations for perturbations about a travelling wave solution $\bar{u}(\xi)$ of (1.1) lead to the eigenvalue problem (1.5) for the linear operator L defined in (1.4). It is well known from the theory of analytic semigroups that if the spectrum of L intersects some half plane $\{\text{Re}\lambda \geq \beta\}$ only at $\lambda = 0$ for $\beta < 0$, and, if $\lambda = 0$ is simple, then the wave $\bar{u}(\xi)$ is stable. More precisely, any solution of (1.1) whose initial data lie uniformly near $\bar{u}(\xi)$ approaches some translate of $\bar{u}(\xi)$ at an exponential rate as $t \to +\infty$; (see (Henry 1981; Bates and Jones 1988)). Furthermore, it can be shown that if K is a simple closed curve in $\{\text{Re}\lambda \geq \beta\}$ which contains $\lambda = 0$ in its interior and which contains a suitably large portion of the unstable half plane

$\lambda \geq 0$, then K contains all potentially unstable eigenvalues of L. This condition on K is independent of the smallness of the entries of the diffusion matrix D (see Proposition 2.2 in (Alexander et al., 1990)). Hence the underlying wave will be stable if there is exactly one eigenvalue of L inside such K, counting multiplicity.

B Hypotheses

Our first assumption is that (1.1) admits a travelling wave $\bar{u}(\xi)$ which tends to limits $u_\pm$ as $\xi \to \pm\infty$ at an exponential rate. The equations (1.5) are then written as a first order system

$$(3.1) \qquad Y' = A(\xi, \lambda) Y$$

where $Y = (p, q)^t \in \mathbf{C}^{2n}$ and

$$A(\xi, \lambda) = \begin{bmatrix} 0 & I \\ D^{-1}(df(\bar{u}(\xi)) - \lambda I) & -\theta D^{-1} \end{bmatrix}$$

The coefficient matrix tends to limits $A_\pm(\lambda)$ as $\xi \to \pm\infty$. We shall call these limits the asymptotic matrices; they play a crucial role in understanding the nonautonomous problem (3.1).

We say that $\lambda \in \mathbf{C}$ *is an eigenvalue of L* if (3.1) admits a uniformly bounded, nontrivial solution. The structure of the matrices $A_\pm(\lambda)$ will clearly be an important factor in determining when this can occur. We therefore assume that the following holds:

(H) There exists a simple connected region $\Omega \subset \mathbf{C}$ such that $A_\pm(\lambda)$ each have k eigenvalues with negative real part and $2n - k$ eigenvalues with positive real part for all $\lambda \in \Omega$, when $1 \leq k \leq 2n - 1$ is the same for $A_-(\lambda)$ and $A_+(\lambda)$.

The continuous spectrum of L is contained between the curves where $A_-(\lambda)$ or $A_+(\lambda)$ has a pure imaginary eigenvalue; thus (H) implies that Ω is disjoint from the continuous spectrum. In particular, if the end states u_- and u_+ of the wave are stable solutions of (1.1), then Ω can be chosen to be a half plane $\mathrm{Re}\lambda > \beta$ for some $\beta < 0$, so that an instability can only be caused by an eigenvalue λ of finite multiplicity with $\mathrm{Re}\lambda > 0$; in this case it can be shown that $k = n$. (see Proposition 2.2 in (Alexander et al., 1990)).

In the following we shall denote the eigenvalues of $A_\pm(\lambda)$ by $\mu_i^\pm(\lambda)$ where $\mathrm{Re}\mu_i^\pm(\lambda) > 0$ (resp. < 0) for $k + 1 \leq i \leq 2n$ (resp. $1 \leq i \leq k$). Let $e_i^\pm(\lambda)$ be eigenvectors associated with $\mu_i^\pm(\lambda)$, and let $S^\pm(\lambda)$ (resp. $U^\pm(\lambda)$) be the span of $e_i^\pm(\lambda)$ for $i \geq k+1$ (resp. $i \leq k$). These subspaces are the stable and unstable subspaces of $A_\pm(\lambda)$.

C The main result

We shall equate the number of eigenvalues counting multiplicity of L inside K with the first Chern number $c_1(\mathcal{E}(K))$ of a certain complex k-plane bundle $\mathcal{E}(K)$ with a base space B which is homeomorphic to a real 2-sphere. $\mathcal{E}(K)$ is called the *augmented unstable bundle*; its construction is a central issue and will be sketched below. The first Chern number of a bundle $\mathcal{E}$ over B is obtained from the first Chern class $c_1[\mathcal{E}] \in H^2(B; \mathbf{Z})$ and applying it to the fundamental class $[B] \in H_2(B; \mathbf{Z})$, i.e.,

$$c_1(\mathcal{E}) = \langle c_1[\mathcal{E}], [B] \rangle;$$

the orientation on B is chosen so that c_1 of the canonical line bundle over B is $+1$ (Milnor and Stasheff 1974). Later, we shall give a more concrete characterization of c_1.

The main intermediary linking $c_1(\mathcal{E}(K))$ to the eigenvalue count is a certain analytic function $D(\lambda)$ called the *Evans function*, whose roots (counting order) coincide with the eigenvalues of L (counting multiplicity). $D(\lambda)$ is essentially the $2n$-fold wedge of certain distinguished solutions of (3.1).

Theorem 3.1 *Suppose that λ is not an eigenvalue of L for all $\lambda \in K$. Then the following three numbers are equal:*

(i) *the first Chern number, $c_1(\mathcal{E}(K))$;*

(ii) *the winding number $W(D(K))$ of the curve $D(K)$ relative to $\lambda = 0$;*

(iii) *the number of eigenvalues of L inside K counting multiplicity.*

Remark. The geometric characterization (i) is particularly useful in applications. For example, if $\mathcal{E}(K)$ can be decomposed as a Whitney sum,

$$\mathcal{E}(K) = \oplus_{i=1}^{\ell} \mathcal{E}_i(K)$$

then the additive property of c_1

$$c_1(\mathcal{E}(K)) = \sum_{i=1}^{\ell} c_1(\mathcal{E}(K))$$

provides a powerful computational tool for decomposing the calculation into substantially simpler pieces. This approach is well suited to singular perturbation problems wherein the underlying wave $\bar{u}(\xi, \varepsilon)$ is resolved into slow segments which are separated by rapid transition layers as a small parameter ε tends to zero. The fast-slow structure is inherited by the linearized equations; this frequently suggests a natural choice for a Whitney sum decomposition of $\mathcal{E}(K, \varepsilon)$.

A second important property of c_1 is homotopy invariance; this accommodates the passage to a singular limit of each summand in a particularly simple manner, since the limits of the bundles $\mathcal{E}_i(K, \varepsilon)$ are frequently well behaved as $\varepsilon \to 0$ even though the associated solutions which span the fibres of the $\mathcal{E}_i$'s become singular. At $\varepsilon = 0$, the governing equations are substantially simpler, and the eigenvalue count, and hence $c_1(\mathcal{E}_i(K))$, can be computed from direct methods, such as Sturm-Liouville theory.

D Construction of $\mathcal{E}(K)$

The hyperbolicity of $A_\pm(\lambda)$ implies that an eigenfunction $Y(\xi,\lambda)$ decays to zero as $\xi \to \pm\infty$. We therefore focus on the sets of solutions of (3.1),

$$\Phi_-(\xi,\lambda) = \{Y(\xi,\lambda) : Y(\xi,\lambda) \to 0 \text{ as } \xi \to -\infty\}$$

$$\Phi_+(\xi,\lambda) = \{Y(\xi,\lambda) : Y(\xi,\lambda) \to 0 \text{ as } \xi \to +\infty\}$$

It follows from (H) that $\Phi_-(\xi,\lambda)$ is a k-dimensional subspace of $\mathbf{C}^{2n}$ and that $\Phi_+(\xi,\lambda)$ is $(2n-k)$-dimensional. This can be seen by augmenting (3.1) with an additional real variable $\tau \in [-1,1]$,

(3.2)
$$Y' = A(\tau,\lambda)Y$$
$$\tau' = \kappa(1-\tau^2), \quad \tau(0) = 0$$

where $0 < \kappa \ll 1$ and $A(\tau,\lambda) = A(\xi(\tau),\lambda)$. Φ_- is then the intersection of the unstable manifold $W^u(\vec{0},-1)$ of the critical point $(\vec{0},-1)$ with $\mathbf{C}^{2n} \times \{\tau(\xi)\}$; Φ_+ is obtained in a similar manner from the stable manifold of $(\vec{0},+1)$. In the following it will be convenient to reparameterize $\Phi_\pm(\tau,\lambda)$ with the compactified variable τ. Φ_- (resp. Φ_+) is a k dimensional (resp. $2n-k$ dimensional) bundle over $[-1,1) \times \Omega$ (resp. $(-1,1] \times \Omega$); Φ_- is the *unstable bundle* and Φ_+ is the *stable bundle*. Clearly, λ is an eigenvalue of L if and only if $\Phi_-(\tau,\lambda) \cap \Phi_+(\tau,\lambda) \neq \{0\}$.

The augmented unstable bundle $\mathcal{E}(K)$ will be constructed from Φ_-. Roughly, it measures the twisting of the k-plane $\Phi_-(\tau,\lambda)$ for λ restricted to K, as τ varies from -1 to $+1$. A dual construction could also be based on Φ_+; the resulting bundle carries the same information as $\mathcal{E}(K)$.

It will be convenient to first mention some elementary facts about bundles. Let $G_k(\mathbf{C}^{2n})$ denote the Grassmannian of k-planes in $\mathbf{C}^{2n}$. $G_k(\mathbf{C}^{2n})$ is a compact manifold obtained by a suitable identification on the Stiefel manifold of k-frames. The canonical bundle $\Gamma_k(\mathbf{C}^{2n})$ is the bundle over $G_k(\mathbf{C}^{2n})$ whose fibre over a point $g \in G_k(\mathbf{C}^{2n})$ is the k-plane associated with g. Given a bundle $\mathcal{E}$, $\pi : E \to B$, there is an induced map $\hat{e} : B \to G_k(\mathbf{C}^{2n})$, namely $\hat{e}(b) = \pi^{-1}(b)$. The pullback $\hat{e}^*$ of $\Gamma_k(\mathbf{C}^{2n})$

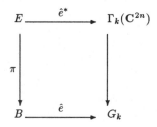

is isomorphic to E. Hence a k-plane bundle over B is equivalent to a continuous map $\hat{e}$ from B into $G_k(\mathbf{C}^{2n})$. Thus if $B \subset B_1$, a continuous extension of $\tilde{e}$ to B_1 continuously extends E to a bundle $\mathcal{E}_1$ over B_1.

The bundle $\mathcal{E}(K)$, $\pi : E \to B$ is defined as follows. The base space B will be the cylinder $B_* = (-1,1) \times K$ together with a left and a right cap, $B_L = \{-1\} \times \overline{K^0}$ and $B_R = \{1\} \times \overline{K^0}$, where K^0 is the interior of K. Thus $B = B_L \cup B_* \cup B_R$ is homeomorphic to S^2. Over the sides, B_*, we define $\pi^{-1}(\tau, \lambda) = \Phi_-(\tau, \lambda)$. In order to see how to extend this to the caps, we need the following lemma.

Lemma 3.2 (i) *For $\lambda \in \overline{K^0}$, $\Phi_-(\tau, \lambda) \to U^-(\lambda)$ as $\tau \to -1$.*
(ii) *Suppose that λ is not an eigenvalue of L for all $\lambda \in K$. Then $\Phi_+(\tau, \lambda) \to U^+(\lambda)$ as $\tau \to +1$.*

The convergence in each case is in $G_k(\mathbf{C}^{2n})$. The defining condition of Φ_- at $\tau = -1$ implies (i), since $U^-(\lambda)$ is the unstable subspace of $A(-1, \lambda)$. The proof of (ii) can be found in (Alexander et al., 1990). The lemma permits us to extend Φ_- to a continuous map from B into $G_k(\mathbf{C}^{2n})$, namely,

$$\Phi_-(\tau, \lambda) = \begin{cases} U^-(\lambda) & (\tau, \lambda) \in B_L \\ \Phi_-(\tau, \lambda) & (\tau, \lambda) \in B_* \\ U^+(\lambda) & (\tau, \lambda) \in B_R \end{cases}$$

The augmented unstable bundle $\mathcal{E}(K)$ is then the pullback $\Phi_-^* \Gamma_k(\mathbf{C}^{2n})$.

Given a k-plane bundle $\mathcal{E} = (E, B, \pi)$ there is an equivalent line bundle $\bigwedge^k \mathcal{E} = (\bigwedge^k E, B, \pi)$ called the *determinant bundle* whose fibres lie in $\bigwedge^k \mathbf{C}^{2n}$. Let i be the Plücker embedding from $G_k(\mathbf{C}^{2n})$ into the projectivization $P(\bigwedge^k \mathbf{C}^{2n})$ of the kth exterior power of $\mathbf{C}^{2n}$. If $\hat{e} : B \to G_k(\mathbf{C}^{2n})$ is the map determined by $\mathcal{E}$ then $\bigwedge^k \mathcal{E}$ is $(i \circ \hat{e})^* \Gamma_1(\bigwedge^k(\mathbf{C}^{2n}))$. This bundle is equivalent to $\mathcal{E}$ and in particular, $c_1(\mathcal{E}) = c_1(\bigwedge^k \mathcal{E})$ (Alexander et al., 1990). It will be more convenient to work with $\bigwedge^k \mathcal{E}$.

E The Evans function

The Evans function, $D(\lambda)$, is constructed as follows. For each $\lambda \in K \cup K^0$ let $\alpha^\pm(\lambda)$ (resp. $\beta^\pm(\lambda)$) be nonzero elements of $\bigwedge^k \mathbf{C}^{2n}$ associated with $U^\pm(\lambda)$ (resp. $S^\pm(\lambda)$), the unstable (resp. stable) subspaces of $A_\pm(\lambda)$ under the Plücker embedding. These can be chosen analytically in λ. The equations (3.2) induce a flow

(3.3)
$$\begin{aligned} Y_j' &= A_j(\tau, \lambda) Y_j \\ \tau' &= \kappa(1 - \tau^2) \end{aligned}$$

on $\bigwedge^j \mathbf{C}^{2n} \times [-1,1]$ for every j. It is easily seen that $\alpha^\pm(\lambda)$ and $\beta^\pm(\lambda)$ are eigenvectors of $A_k(\pm 1, \lambda)$ and $A_{2n-k}(\pm 1, \lambda)$ associated with eigenvalues

$$a^\pm(\lambda) = \sum_{i=1}^{k} \mu_i^-(\lambda) \text{ and } b^\pm(\lambda) = \sum_{i=k+1}^{2n} \mu_i^+(\lambda).$$

It follows that $(\alpha^-(\lambda), -1)$ and $(\beta^+(\lambda), +1)$ are critical points of the rescaled systems

$$\eta_k' = [A_k(\tau, \lambda) - a^-(\lambda)I]\eta_k \quad \text{and}$$

$$\eta_{2n-k}' = [A_{2n-k}(\tau, \lambda) - b^+(\lambda)I]\eta_{2n-k}$$

$$\tau' = \kappa(1 - \tau^2).$$

By applying the centre unstable and centre stable manifold theorems to these flows we obtain solutions $\eta_k(\tau, \lambda)$ and $\eta_{2n-k}(\tau, \lambda)$ which satisfy

(3.4)
$$\eta_k(\tau, \lambda) \to \alpha^-(\lambda) \quad \text{as} \quad \tau \to -1$$
$$\eta_{2n-k}(\tau, \lambda) \to \beta^+(\lambda) \quad \text{as} \quad \tau \to +1.$$

Next, we rescale these forms to recover solutions (3.3):

$$Y_k(\tau, \lambda) = \eta_k(\tau, \lambda) \exp(-a^-(\lambda)\xi(\tau))$$
$$Y_{2n-k}(\tau, \lambda) = \eta_{2n-k}(\tau, \lambda) \exp(-b^+(\lambda)\xi(\tau)).$$

Note that Y_k and Y_{2n-k} are nonzero sections of $\bigwedge^k \Phi_-$ and $\bigwedge^{2n-k} \Phi_+$, respectively. $D(\lambda)$ is defined to be

(3.5) $$D(\lambda) = Y_k(\tau, \lambda) \wedge Y_{2n-k}(\tau, \lambda) \exp(-\int_0^{\xi(\tau)} \text{tr} A(s, \lambda) ds).$$

Noting that $Y_k \wedge Y_{2n-k}$ is a solution of (3.3) with $j = 2n$ it follows immediately from Abel's formula that $D(\lambda)$ is independent of τ. We mention two other crucial properties of $D(\lambda)$.

Lemma 3.3 (i) $D(\lambda)$ *is analytic in* λ.
(ii) *The roots* λ *of* $D(\lambda)$ *(counting order) coincide with the eigenvalues of* L *(counting algebraic multiplicity)*.

The proof of (i) and (ii) are somewhat involved (see(Alexander et al., 1990)). A more direct proof of (ii) has recently been obtained by Gardner and Jones (1990b) for boundary value problems.

F The proof of theorem 3.1

The final step is to relate $c_1(\bigwedge^k \mathcal{E}(K))$ to $D(\lambda)$. To this end we give an alternative characterization of c_1 for line bundles $\mathcal{E} = (E, S^2, \pi)$ over S^2. If $S^2 = S^- \cup S^+$ is expressed as the union of two hemispheres with $S^+ \cap S^- = S^1$, there exist trivializations

$$\chi_\pm : S^\pm \times \mathbf{C} \to E|S^\pm.$$

The composition $\chi_+^{-1} \circ \chi_- : S^1 \times \mathbf{C} \to S^1 \times \mathbf{C}$ induces a nonzero map $f : S^1 \to \mathbf{C}-\{0\}$ given by

$$\chi_+^{-1} \circ \chi_-(\lambda, e) = (\lambda, f(\lambda)e).$$

The map f, which is called the *gluing map* of the bundle, represents a class in $\pi_1(\mathbf{C} - \{0\}) = \pi_1(S^1)$, so that $[f] = n(f)[i]$ where $i : S^1 \to \mathbf{C} - \{0\}$ is the identity and $n(f) \in \mathbf{Z}$. It then follows that $c_1(\mathcal{E}) = n(f)$ (see (Alexander et al., 1990)).

Next, consider the line bundle $\bigwedge^k \mathcal{E}(K)$. We take $S^- = B_L \cup \bar{B}_*$ and $S^+ = B_R$, so that $S^- \cap S^+ = \{+1\} \times K$. Trivializations for the restrictions to $S^\pm$ are obtained as follows. For S^+, set $b = (+1, \lambda)$ and

$$\chi_+(b, z) = (b, z\alpha_+(\lambda)).$$

For S^- it is tempting to use $\eta_k(\tau, \lambda)$ since by (3.4), $\eta_k(\tau, \lambda)$ matches up with the trivialization $\alpha_-(\lambda)$ over B_L. However, $\eta_k(\tau, \lambda)$ necessarily blows up as $\tau \to +1$. Thus it is necessary to rescale $\eta_k(\tau, \lambda)$ by defining

$$\tilde{\eta}_k(\tau, \lambda) = \begin{cases} \alpha_-(\lambda), & \tau = -1, \quad \lambda \in K^0 \\ \eta_k(\tau, \lambda), & -1 \leq \tau \leq 0, \quad \lambda \in K \\ \eta_k(\tau, \lambda)\rho(\tau, \lambda), & 0 \leq \tau < 1, \quad \lambda \in K \end{cases}$$

where

$$\rho(\tau, \lambda) = \exp[(-a^-(\lambda) - a^+(\lambda))\xi(\tau)].$$

It then follows that $\tilde{\eta}_k(\tau, \lambda)$ tends uniformly to a finite, continuous limit $\zeta(\lambda)$ as $\tau \to +1$, for $\lambda \in K$; see Lemma 6.1 in (Alexander et al., 1990). Thus for $b \in B_L \cup \bar{B}_*$ set

$$\chi_-(b, z) = \begin{cases} (b, z\alpha_-(\lambda)), & b \in B_L \\ (b, z\tilde{\eta}_k(b)), & b \in B_* \\ (b, z\zeta(\lambda)), & b \in \{+1\} \times K. \end{cases}$$

Since $\zeta(\lambda)$ and $\alpha^+(\lambda)$ are both nonzero sections of the restriction of $\bigwedge^k \mathcal{E}(K)$ to $\{+1\} \times K$ it follows that $\zeta(\lambda) = f(\lambda)\alpha^+(\lambda)$ for some nonzero function $f : K \to \mathbf{C} - \{0\}$. It is easily seen that $f(\lambda)$ is the gluing map for the bundle induced by $\chi_+^{-1} \circ \chi_-$.

The proof is completed by noting that the limit of the expression in (3.5) for $D(\lambda)$ exists as $\tau \to +1$, as can be seen by redistributing the exponential weight appropriately in each factor. The limit is then seen to be

$$\begin{aligned} D(\lambda) &= \zeta(\lambda) \wedge \beta^+(\lambda) \\ &= f(\lambda)\alpha^+(\lambda) \wedge \beta^+(\lambda) \\ &= f(\lambda)d(\lambda)\bar{1}, \end{aligned}$$

where $d(\lambda)$ is a *nonzero* analytic function on $\overline{K^0}$, and $\bar{1}$ is the fundamental volume form. It then follows that $W(D(K)) = W(f(K)) = [f]$, completing the proof.

G An application

The stability of the travelling waves of Theorem 2.4 for small ε was established by Gardner and Jones (1990a) using the above machinery. It is not possible to present more than a brief account of this work here.

Theorem 3.4 *In addition to the hypotheses of Theorem 2.4 suppose that if Z_R is the right branch of $f_2(u) = 0$, then*

$$(H_2): \qquad \max{}_{u \,\in\, Z_R} \frac{|f_{1u_2}(u) f_{2u_1}(u)|}{f_{1u_1}(u)^2} < 1.$$

Then the travelling waves of Theorem 2.4 are stable for all sufficiently small $\varepsilon > 0$.

Since O and C are stable rest points, it follows that for the region $\Omega = \{\text{Re}\lambda > \beta\}$ for some $\beta < 0$ the asymptotic (4×4) matrices $A_\pm(\lambda)$ have a $2-2$ splitting, i.e., $k = 2$ for $\lambda \in \Omega$. This enables us to construct the (two-dimensional) augmented unstable bundle $\mathcal{E}(K)$ for simple closed curves $K \subset \Omega$ disjoint from the spectrum of L. We choose K to be a curve containing $\lambda = 0$ in its interior which contains a suitably large portion of $\text{Re}\lambda \geq 0$.

The goal of the proof is to show that $c_1(\mathcal{E}(K)) = 1$. To this end, several issues must be addressed, namely (1) the location of suitable closed curves K disjoint from the spectrum of L, (2) the decomposition of $\mathcal{E}(K)$ into a Whitney sum $\mathcal{E}_1(K) \oplus \mathcal{E}_2(K)$ of line bundles, and (3), the computation of $c_1(\mathcal{E}_1(K))$ and $c_1(\mathcal{E}_2(K))$. These issues are simultaneously resolved by considering the projectivized flow induced by (3.2) on $\mathbf{C}P^3 \times [-1,1] = G_1(\mathbf{C}^4) \times [-1,1]$, namely

$$(3.6) \qquad \begin{aligned} \hat{Y}' &= \hat{A}(\hat{Y}, \tau, \lambda) \\ \tau' &= \kappa(1 - \tau^2). \end{aligned}$$

The critical points of (3.6) are $(\hat{e}_i^\pm(\lambda), \pm 1)$, $1 \leq i \leq 4$, where $e_i^\pm(\lambda)$ are the eigenvectors of $A_\pm(\lambda)$. The unstable subspace $U^-(\lambda)$ is spanned by $e_1^-(\lambda)$ and $e_2^-(\lambda)$, where $e_1^-(\lambda)$ is associated with an eigenvalue of order 1 and $e_2^-(\lambda)$ is associated with an eigenvalue of order ε. A solution of (3.6) determines a one dimensional subbundle of

$\mathcal{E}(K)$ (see section 3.D) provided that the solutions behaviour can be characterized at $\tau = \pm 1$. The goal is to produce solutions $e_i(\tau, \lambda)$, $i = 1, 2$, such that

$$\hat{e}_i(\tau, \lambda) \to \hat{e}_i^{\pm}(\lambda) \text{ as } \tau \to \pm 1.$$

This is a connecting orbit problem for the flow (3.6) on $\mathbb{C}P^3 \times [-1, 1]$. The fast solution, $\hat{e}_1(\tau, \lambda)$, is determined by the behaviour of the underlying wave in the transition layer while the slow solution, $\hat{e}_2(\tau, \lambda)$, is determined by the behaviour of the wave in the slow manifolds. In each case the relevant solution is approximated by setting $\varepsilon = 0$ in fast and slow linearizations of (2.9) and (2.10). An advantage of our geometric approach is that these singular limits are particularly easy to identify on the level of the bundle, since the *fibres* of the bundle are well behaved as $\varepsilon \to 0$ while associated sections arising as solutions of the linearized equations become singular. Hence the geometric approach using $\mathcal{E}(K)$ enables us to identify the correct reduced eigenvalue problems at $\varepsilon = 0$. Furthermore, the topological nature of the Chern number accommodates the passage to the singular limit in that spectral information contained in the fast and slow reduced systems at $\varepsilon = 0$ continues to $\varepsilon > 0$. In particular, this enables us to compute $c_1(\mathcal{E}_1(K))$ and $c_1(\mathcal{E}_2(K))$ for $\varepsilon > 0$. The fast reduced problem for $\mathcal{E}_1(K)$ turns out to be the linearized problem for the travelling wave for the scalar bistable equation considered in section 2.D. This wave is known to be stable (Fife and Mcleod 1978), hence $c_1(\mathcal{E}_1(K)) = 1$. The slow reduced system for $\mathcal{E}_2(K)$ is a nonstandard eigenvalue problem because the eigenvalue parameter λ enters into the equations in a nonlinear manner. However the eigenvalue problem for this problem can be determined from Sturm-Liouville theory. It turns out that with hypothesis (H2), this eigenvalue problem has no eigenvalues for $\text{Re}\lambda > \beta$ for some $\beta < 0$; hence $c_1(\mathcal{E}_2(K)) = 0$, completing the proof.

References

[1] Alexander, J, Gardner, R, and Jones, C. (1990). A topological invariant arising in the stability analysis of travelling waves. To appear in *Journal für Reine und Angewandte Mathematik*.

[2] Bates, P. and Jones, C. Invariant manifolds for semilinear partial differential equations. In press.

[3] Conley, C. (1978). *Isolated invariant sets and the generalized Morse index.* CBMS regional conference series in Applied Math., **38**. A.M.S., Providence, R. I.

[4] Conley, C. and Gardner, R. (1984). An application of the generalized Morse index to travelling wave solutions of a competitive reaction-diffusion model. *Indiana University Mathematics Journal*, **33**, 319-43.

[5] Conley, C. and Smoller, J. (1978). Isolated invariant sets of parameterized systems of partial differential equations. In *The structure of attractors in dy-*

namical systems, (eds N. Markley, J. C. Martin, and W. Perizo), Lecture Notes in Mathematics, **668**. Springer-Verlag, Berlin.

[6] Fife, P. and Mcleod, J. B. (1978). The approach of solutions of nonlinear diffusion equations to travelling front solutions. *Archive for Rational Mechanics and Analysis*, **65**, 335-61.

[7] Gardner, R. (1984). The existence of travelling wave solutions of predator-prey systems via the connection index. *SIAM Journal on Applied Mathematics*, **44**, 56-79.

[8] Gardner, R. and Jones, C. (1990a). Stability of travelling wave solutions of predator prey systems. To appear in *Transactions of the American Mathematical Society*.

[9] Gardner, R. and Jones, C. (1990b). A stability index for steady state solutions of boundary value problems for parabolic systems. To appear in *Journal of Differential Equations*.

[10] Henry, D. (1981). *The geometric theory of semilinear parabolic equations*. Lecture notes in Mathematics, **840**. Springer-Verlag, Berlin.

[11] Milnor, J. and Stasheff, J. D. (1974). *Characteristic classes*. Annals of Mathematical Studies, **76**. Princeton N.J.

[12] Nishiura, Y. and Fujii, H. (1987). Stability of singularly perturbed solutions to systems of reaction-diffusion equations. *SIAM Journal on Mathematical Analysis*, **18**, 1726-70.

[13] Nishiura, Y, Mimura, M., Ikeda, H., and Fujii, H. Singular limit analysis of stability of travelling wave solutions in bistable reaction-diffusion equations. In press.

[14] Terman, D. Stability of planar wave solutions to a combustion model. In press.

Maximum principles and decoupling for positive solutions of reaction-diffusion systems

Jesús Hernández [*]
Universidad Autónoma

0 Introduction

Reaction-diffusion systems have been intensively studied during recent years; see (Smoller 1983), where many references can be found. In particular, existence and multiplicity of solutions for the corresponding elliptic systems have been considered, mostly for the case of positive solutions. Indeed, for many systems arising in applications, they are the only physically meaningful solutions (concentrations, populations, etc).

We shall survey some recent results concerning existence and multiplicity of positive solutions for some classes of steady-state reaction-diffusion systems, obtained by using decoupling techniques together with a variety of topological and analytical methods (sub and supersolutions, global bifurcation, continuation, critical point theory, etc). Needless to say, this only covers a small part of the results currently available for positive solutions of elliptic reaction-diffusion systems. But, some limitation seems to be necessary in order to keep this account at a reasonable length. There are, however, interesting, and difficult, open problems arising from other types of examples.

The decoupling technique, which we will describe in more detail later, consists in reducing a system of two equations to a single nonlinear equation containing a nonlocal, linear or nonlinear, term. The method has some obvious shortcomings (for example, it is very difficult to apply to systems of three or more equations), but it has been used with some success in a number of systems arising in applications (e.g., morphogenesis, population dynamics, Fitzhugh-Nagumo type systems) raising interesting open problems concerning the global structure of the bifurcation diagrams for positive solutions and related topics. In particular, we will emphasize the role played by the Maximum Principle (or its lack) concerning positivity.

Decoupling for a class of systems is described in section 1. This technique was introduced by Rothe(1981), Lazer and McKenna (1982), and Brown (1983), and has been used thereafter by many authors. Roughly speaking, these systems may be reduced to a nonlinear equation

$$-\Delta u + Bu = g(u) \quad \text{in } \Omega$$

$$u = 0 \quad \text{on } \partial\Omega,$$

---

[*]This work was partially supported by Project 3308/83 of CAICYT, Spain

where B is the (integral) solution operator for a linear equation and g is a given nonlinearity. This enables the problem to be considered as a nonlocal linear perturbation of the nonlinear equation $-\Delta u = g(u)$ for which many results are available (see the surveys (Amann 1976a; Lions 1982); see also (Amann and Laetsch 1976; Ambrosetti and Rabinowitz 1973; Berestycki 1981; Brezis and Nirenberg 1983; Brezis and Turner 1977; Hernández 1986; Rabinowitz 1971a,b,c; Sattinger 1972; Smoller 1983) and the corresponding bibliographies). This method may be applied if , when fixing one of the unknowns, the resulting equation is linear or, more generally, has a unique solution. The above interpretation suggests trying to extend to the perturbed problem the results already known when $B \equiv 0$.

The basic tool for most of the results is an interesting Maximum Principle due to de Figueiredo and Mitidieri (1986), which gives rise to some intriguing open problems, see Remarks 1.1-1.5, 2.3, 3.2 and 4.2 below, (Caristi and Mitidieri 1988), and (de Figueiredo and Mitidieri 1988). For an interesting related (but not equivalent!) problem, namely the existence of eigenvalues with positive eigenfunctions for elliptic systems, see the work by Hess (1983), Cantrell and Schmidt (1986), and the survey by Cosner (1988).

Sections 2 and 3 are devoted to the case $g(u) = \lambda u - f(u)$, with f asymptotically linear. To the best of our knowledge, this is the only case where a complete description of the bifurcation diagram for positive solutions can be given and this is the reason why we consider first this problem. Roughly speaking, the results for $B \equiv 0$ are still true under some additional assumptions (for $B \equiv 0$ see (Sattinger 1972; Hernández 1977, 1986; Berestycki 1981)). In section 2 we study the case of f convex by using sub and supersolutions and in section 3 we give an alternative proof by using a local inversion and continuation method due to Rabinowitz (1971a); the latter is also used to treat the case of f concave, where sub and supersolutions are not available. These results are taken from (Hernández 1988), some being announced in (Hernández 1987).

The sublinear case is considered in section 4. This problem was studied first in (Rothe 1981) (see also(Rothe and de Mottoni 1979)) for f odd by using critical point theory, and then in (Lazer and McKenna 1982) for f not necessarily odd by using degree theory, but no information is given about the positivity of the solutions. First, we include a general existence result obtained by sub and supersolutions (de Figueiredo and Mitidieri 1986) and then we pass to the case $g(u) = \lambda u - u^3$. Curiously enough, sub and supersolutions (de Figueiredo and Mitidieri 1986) and continuation (Hernández 1988, 1987) give different bounded intervals for existence and uniqueness of positive solutions; the difficulties which arise come, once again, from the Maximum Principle. Existence is proved by using global bifurcation together with *a priori* estimates for positive solutions, but uniqueness remains open. Some observations concerning the superlinear case are also included (see (de Figueiredo and Mitidieri 1986); Sobreira 1985 a, b)).

The sublinear case is considered again in section 5, but this time by using critical point theory, namely the *Mountain Pass Lemma* (Ambrosetti and Rabinowitz 1973). Here the main results are due to Klaasen and Mitidieri (1986), Klaasen (1987) and de Figueiredo and Mitidieri (1986). We give the results of the latter, including the necessary *a priori* estimates . A more difficult problem, a system with nonlinear decoupling and "critical exponent" treated by Mancini and Mitidieri (1987) is also included. These results are in some sense an extension to systems of the work by Brezis and Nirenberg (1983).

Finally, section 6 is devoted to some more involved problems arising in population dynamics where both equations are nonlinear. The main example is a predator-prey model considered by Conway *et al.* (1982), but some questions concerning existence and uniqueness for some range of the parameters remained open. Existence was settled independently by Blat and Brown (1984) (see also (Blat 1987; Brown 1983, 1987) for related results) by using nonlinear decoupling and global bifurcation and by Dancer (1984), this time by using degree theory in cones. In this case the methods of the previous sections do not work : a Maximum Principle allowing the use of sub and supersolutions is not available, and continuation cannot be applied because of lack of smoothness. For the sake of brevity we only include the results of (Blat and Brown 1984), where the competition system is also treated. An interesting variant is the Holling-Tanner model (Blat and Brown 1986a). Uniqueness seems to be an interesting open problem; to the best of our knowledge there is only a proof by Dancer (1985) in the one-dimensional case. We would like to emphasize that the techniques employed by Dancer (1984, 1985) also apply to more general problems and give interesting stability and multiplicity results. Other stability results are due to Blat and Brown (1984, 1986b), Cantrell and Cosner (1987), Cosner and Lazer (1984), Lazer and McKenna (1982), Rothe and Mottoni (1979).

The author would like to thank S. Angenent, J. Blat and E. Mitidieri for some very useful conversations.

1 Linear decoupling: the Maximum Principle

We are interested in the study of the semilinear elliptic system

(1.1)
$$-\Delta u = g(u) - v \quad \text{in } \Omega$$
$$-\Delta v = \delta u - \gamma v \quad \text{in } \Omega$$
$$u = v = 0 \quad \text{on } \partial\Omega$$

where Ω is a bounded domain in $\mathbf{R}^N$ ($N \geq 1$), with a very smooth boundary $\partial\Omega$ (in particular we assume that it is $C^{2,\alpha}$ for every $0 < \alpha < 1$), $\delta > 0$ and $\gamma > 0$ are given real numbers, and g is a C^1 function satisfying various sets of additional assumptions. A system of this type arises in the study of morphogenesis (Rothe 1981; Lazer and McKenna 1982).

The system (1.1) can be reduced to a single nonlinear equation by using a decoupling technique (Rothe 1981; Lazer and McKenna 1982; Brown 1983). More precisely, for any $u \in L^2(\Omega)$ the linear equation

$$-\Delta v + \gamma v = \delta u \quad \text{in } \Omega$$
$$v = 0 \quad \text{on } \partial\Omega$$

possesses a unique solution $v = Bu \in H^2(\Omega)$. (We employ the usual notation $C^{k,\alpha}(\overline{\Omega})$, $0 < \alpha < 1$, for the spaces of Hölder continuous functions, and $H^m(\Omega)$ for Sobolev spaces on $L^2(\Omega)$). It is well-known that $B : L^2(\Omega) \to H^2(\Omega)$, $B : C^\alpha(\overline{\Omega}) \to C^{2,\alpha}(\overline{\Omega})$, etc., are well-defined, compact linear operators in $L^2(\Omega)$, $C^\alpha(\overline{\Omega})$, etc. Moreover, by the usual Maximum Principle, if $u \in C(\overline{\Omega})$ and $u \geq 0$, then $Bu \equiv 0$ or $Bu > 0$ in Ω. The system (1.1) is equivalent to the nonlinear equation

(1.2)
$$-\Delta u + Bu = g(u) \quad \text{in } \Omega$$
$$u = 0 \quad \text{on } \partial\Omega,$$

where the nonlocal integral linear operator $B : L^2(\Omega) \to L^2(\Omega)$ is defined as above. Now it is very easy to pass from solutions of (1.2) to solutions of (1.1).

We consider the operator $T = -\Delta + B$ with domain $D(T) = H^2(\Omega) \cap H_0^1(\Omega)$ in $L^2(\Omega)$. T is a positive, symmetric, closed operator. We need some spectral theory for T (see the above mentioned papers and (de Figueiredo and Mitidieri 1986)). The first lemma is a comparison result for eigenvalues extending the classical theory (Courant and Hilbert 1953).

Lemma 1.1 *Let $\rho \in C(\overline{\Omega})$. There exists an increasing sequence $\gamma_n(\rho)$, $n = 1, 2, \ldots$ of real numbers, where the $\gamma_n(\rho)$ are counted according to multiplicity, such that $\gamma_n(\rho) \to +\infty$ as $n \to +\infty$ and the problem*

(1.3)
$$-\Delta w + Bw + \rho(x)w = \gamma w \quad \text{in } \Omega$$
$$w = 0 \quad \text{on } \partial\Omega,$$

has a nontrivial solution if and only if $\gamma = \gamma_k(\rho)$ for some k. Moreover, if $\rho(x) \leq \bar{\rho}(x)$, $\rho \not\equiv \bar{\rho}$, with $\bar{\rho} \in C(\overline{\Omega})$, then $\gamma_n(\rho) < \gamma_n(\bar{\rho})$.

We denote by $\hat{\lambda}_k(\rho)$ the eigenvalues of (1.3) for $\rho \in C(\overline{\Omega})$; for $\rho \equiv 0$ we write $\hat{\lambda}_k \equiv \hat{\lambda}_k(0)$. If $B \equiv 0$ we use the notation $\lambda_k(\rho)$ and $\lambda_k \equiv \lambda_k(0)$.

Assume that φ_k are the eigenfunctions for the eigenvalues λ_k of $-\Delta$, normalized by $\|\varphi_k\|_\infty = 1$. We have

(1.4)
$$-\Delta \varphi_k = \lambda_k \varphi_k \quad \text{in } \Omega$$
$$\varphi_k = 0 \quad \text{on } \partial\Omega.$$

It is well-known that it is possible to choose $\varphi_1 > 0$ in Ω and $\frac{\partial \varphi_1}{\partial n} < 0$ on $\partial\Omega$.

It is very easy to check that

$$T\varphi_k = \hat{\lambda}_k \varphi_k,$$

where
(1.5)
$$\hat{\lambda}_k = \lambda_k + \frac{\delta}{\gamma + \lambda_k}.$$

The eigenfunctions (φ_k) form a total orthonormal system in $L^2(\Omega)$ and this implies that the $(\hat{\lambda}_k)$ are the only eigenvalues of T. It is not difficult to show that the spectrum $\sigma(T)$ consists only of these eigenvalues.

It is important to note that the sequence $\hat{\lambda}_k$ is not necessarily increasing. A sufficient condition for this sequence to be increasing is

(1.6)
$$\lambda_1 + \gamma > \sqrt{\delta}.$$

In particular, (1.6) implies that the first eigenvalue $\hat{\lambda}_1$ is simple with positive eigenfunction φ_1.

Lemma 1.2 *For any μ in the resolvent of T, $T_\mu = (T - \mu I)^{-1}$ is compact.*

The following theorem is an important tool for the proof of many of the results discussed below.

Theorem 1.1 *(Maximum Principle (de Figueiredo and Mitidieri 1986)). Suppose that (1.6) is satisfied. For any μ such that*

(1.7)
$$2\sqrt{\delta} - \gamma \leq \mu < \hat{\lambda}_1,$$

$u \geq 0$ *implies* $T_\mu u \geq 0$. *Moreover, if* $u \in C(\overline{\Omega})$, $u \geq 0$, $u \not\equiv 0$, *then* $T_\mu u > 0$ *in* Ω *and* $\dfrac{\partial T_\mu u}{\partial n} < 0$ *on* $\partial \Omega$.

We now make a series of remarks on some extensions, comments and open problems related with this interesting result.

Remark 1.1 It is easily seen that $\mu < \hat{\lambda}_1$ is a necessary condition for a Maximum Principle. It is not *a priori* obvious, however, that $2\sqrt{\delta} - \gamma$ is the best lower bound for μ. It has been proved (Sobreira 1986) that the theorem cannot be true for every $\mu < \hat{\lambda}_1$. By using a result for semigroups (Martin 1976) which says that if A is the infinitesimal generator of a semigroup $T(t)$ in a Banach space ordered by a positive cone K, then K is invariant under $T(t)$ if and only if the resolvent $(\lambda I - A)^{-1}$ is positive for λ sufficiently large, it suffices to exhibit initial data such that the solution of the corresponding evolution problem is not always positive.

Remark 1.2 A completely different proof of Theorem 1.1 can be given by using the notion of absolutely monotone functions and the associated integral representation formula by means of a positive Borel measure (see, e.g., (Hirsch 1972)). We are indebted to S. Angenent (personal communication) for this observation.

Remark 1.3 Both proofs of Theorem 1.1 use in an apparently essential way the fact that μ is a constant and do not extend to the operator $-\Delta + B + a(x)I$ for nonconstant $a(x)$. Results for variable coefficients have been obtained recently (de Figueiredo and Mitidieri 1988) by a different method, which consists of embedding a non-cooperative 2×2 system into a 3×3 cooperative one (see also (Mancini and Mitidieri 1987)). However, these results are not strong enough to settle the uniqueness problem left open in section 4, see Remarks 3.2 and 4.2.

Remark 1.4 The nonlocal operator B in (1.2) may be replaced by a more general linear, or even nonlinear, operator; for instance, B could be replaced by the solution operator C of some equation with a unique solution. Roughly speaking, C (and $(-\Delta + C)^{-1}$) is compact and the main problem is to find a Maximum Principle, or some comparison result for $(-\Delta + C)^{-1}$. Moreover, both proofs of Theorem 1.1 rely heavily on the fact that the differential operator $(-\Delta)$ is the same in both equations, and the case of two different uniformly elliptic linear differential operators is still open. (See, however, Theorem 1.2 in (de Figueiredo and Mitidieri 1988). The main results, however, in that paper, which include interesting extensions of well-known classical results (Protter and Weinberger 1967) are stated for the same differential operator). These observations and Remark 1.2 show the interest of the study of $-\Delta + C$ for various C's. Some existence theorems for a particular nonlinear C are obtained in (Hernández 1988).

Remark 1.5 Theorem 1.1 and , more generally, the Maximum Principle for systems, are closedly related with the possibility of decoupling the system (de Figueiredo and Mitidieri 1986, Remark 1.7 and Proposition 1.5; Weinberger 1988; de Figueiredo and Mitidieri 1988). Note that our system (1.1) is not cooperative and that the matrix

$$\begin{pmatrix} \mu & -1 \\ \delta & -\gamma \end{pmatrix}$$

has two distinct real eigenvalues if and only if $\mu > -\gamma + 2\sqrt{\delta}$, which is precisely the bound in (1.7). This fact together with some heuristic considerations seems to suggest the optimality of the interval (1.7), but sharper results recently obtained (Caristi and Mitidieri 1988) for both the one-dimensional and the radial case show that this is not true. Some related problems are considered in (Sweers 1988).

2 The asymptotically linear case: sub and supersolutions

We consider the elliptic system

(2.1)
$$-\Delta u = \lambda u - f(u) - v \quad \text{in } \Omega$$
$$-\Delta v = \delta u - \gamma v \quad \text{in } \Omega$$
$$u = v = 0 \quad \text{on } \partial\Omega,$$

where λ is a real parameter and $f : \mathbf{R} \to \mathbf{R}$ satisfies

(2.2) $\qquad\qquad\qquad f$ is C^2, increasing, $f(0) = 0$,

(2.3) $\qquad\qquad$ the function $\dfrac{f(u)}{u}$ is strictly increasing for $u > 0$
and strictly decreasing for $u < 0$,

(2.4) $$\lim_{u \to +\infty} \frac{f(u)}{u} = f'(\infty) < +\infty,$$

It is clear that $f'(0) < f'(\infty)$. It is possible to take $f'(0) = 0$, without loss of generality, but we prefer to emphasize the role played by both "derivatives" $f'(0)$ and $f'(\infty)$.

According to the preceding section, (2.1) is equivalent to the nonlinear equation

(2.5)
$$-\Delta u + Bu + f(u) = \lambda u \quad \text{in } \Omega$$
$$u = 0 \quad \text{on } \partial\Omega.$$

A first auxiliary result indicates for what values of the parameter λ solutions of (2.5) may exist.

Lemma 2.1 *Assume that f satisfies (2.2)–(2.4). If $u \in C(\overline{\Omega})$, $u \geq 0$ is a nontrivial solution of (2.5), then $\hat{\lambda}_1 + f'(0) < \lambda < \hat{\lambda}_1 + f'(\infty)$.*

Proof (Sketch) Multiply (2.5) by φ_1 and integrate over Ω by using Green's Formula. The result follows from (2.3), (2.4) and Lemma 1.1.

Lemma 2.2 *Assume that $u \in C(\overline{\Omega})$, $u \geq 0$ is a nontrivial solution of (2.5), where (1.6) and (2.2)–(2.4) are satisfied and that*

(2.6) $$f'(\infty) - f'(0) \leq \hat{\lambda}_1 + \gamma - 2\sqrt{\delta}.$$

Then $u > 0$ in Ω and $\dfrac{\partial u}{\partial n} < 0$ on $\partial\Omega$.

Proof (Sketch) By Theorem 1.1 it is enough to find a $\bar{\mu}$ such that, if we rewrite (2.5) as

(2.7)
$$-\Delta u + Bu - \bar{\mu}u = \lambda u - f(u) - \bar{\mu}u \quad \text{in } \Omega$$
$$u = 0 \quad \text{on } \partial\Omega,$$

the Maximum Principle holds for $-\Delta + B - \bar{\mu}I$ and the right-hand side in (2.7) is increasing for $u \geq 0$. This yields condition (2.6).

Remark 2.1 The Maximum Principle seems to be necessary to show that nontrivial positive solutions are strictly positive on Ω. If not, "dead core" phenomena (see, e.g., (Diaz 1985)) cannot be *a priori* excluded.

Theorem 2.1 *Assume that* (1.6), (2.2)–(2.4) *and* (2.6) *are satisfied. Then for any* λ *such that* $\hat{\lambda}_1 + f'(0) < \lambda < \hat{\lambda}_1 + f'(\infty)$ *there exists a unique (classical) nontrivial positive solution* $u(\lambda)$ *for* (2.5). *The mapping* $\lambda \to u(\lambda)$ *from* $(\hat{\lambda}_1 + f'(0), \hat{\lambda}_1 + f'(\infty))$ *into* $C^\alpha(\overline{\Omega})$ *is continuous and increasing.*

Proof (Sketch) It follows from Lemma 1.2 and the proof of Lemma 2.2 that the method of sub and supersolutions (Amann 1976 a) can be used. It is easy to check that $u_0 \equiv c\varphi_1$, for $c > 0$ sufficiently small, is a subsolution for $\hat{\lambda}_1 + f'(0) < \lambda$.

A supersolution $u^0 > u_0$ can be found by exploiting the fact that $\hat{\lambda}_1$ depends continuously and monotonically on the domain Ω. This last fact follows from well-known classical results (Courant and Hilbert 1953), (1.5) and (1.6). Uniqueness can be proved as for $B \equiv 0$ (Hernández 1986).

Remark 2.2 The main difficulty in proving Theorem 2.1 is reformulating the problem in such a way that the method of sub and supersolutions can be applied. The Maximum Principle is vital in this respect.

Remark 2.3 We point out that existence depends (through the supersolution u^0) on the explicit formula (1.5), and this is another difficulty in extending the method to the case of two different differential operators (see Remark 1.4).

Remark 2.4 Assumptions (1.6) and (2.6) are satisfied if, for example, we take γ large for fixed δ.

Remark 2.5 Another asymptotically linear problem, but of a somewhat different type, appears in the study of the spread of bacterial infections (Capasso and Maddalena 1982; Blat and Brown 1986 b). The corresponding elliptic system is

$$-d_1 \Delta u = -a_1 u + bv \quad \text{in } \Omega$$

$$-d_2 \Delta v = -a_2 v + f(u) \quad \text{in } \Omega$$

together with two different boundary conditions of the third kind. Here d_1, d_2, a_1, $a_2 > 0$ are given real numbers, b is a real bifurcation parameter and f satisfies (2.2) and (2.4).

The system can be decoupled by solving the first equation for fixed v, obtaining $u = bK_1(v)$, with $K_1 = (-d_1 \Delta + a_1 I)^{-1}$. Hence we get for v the nonlinear equation

$$-d_2 \Delta v + a_2 v = f(bK_1(v)) \quad \text{in } \Omega$$

which is quite different from (2.5), the nonlocal term arising as $f(bK_1(v))$. The above equation can be reformulated as $u = K_2(F(bK_1(v)))$, where $K_2 = (-d_2\Delta + a_2 I)^{-1}$ and F is the Nemitskii operator associated with f. If $\frac{f(u)}{u}$ is bounded, existence of positive solutions for some b's can be proved by using global bifurcation arguments (Blat and Brown 1986 b). If $\frac{f(u)}{u}$ is also decreasing for $u \geq 0$, then nontrivial positive solutions are unique and the parameter interval can be precisely determined, as in Lemma 2.1.

Remark 2.6 Some multiplicity results for a variant of the system in Remark 2.5 are given in Mirenghi (1987) by using respectively, sub and supersolutions and critical point theory.

3 The asymptotically linear case: continuation

We consider the same problem as in section 2, but this time by using a local inversion theorem of Crandall and Rabinowitz and a continuation argument involving the Implicit Function Theorem.

Suppose f satisfies the assumptions in Theorem 2.1 and $f \in C^3$. We define a mapping $F : \mathbf{R} \times C_0^{2,\alpha}(\overline{\Omega}) \to C^\alpha(\overline{\Omega})$, where $C_0^{2,\alpha}(\overline{\Omega}) = \{u \in C^{2,\alpha}(\overline{\Omega}) | u = 0 \text{ on } \partial\Omega\}$ by

$$F(\lambda, u) = -\Delta u + Bu + f(u) - \lambda u.$$

The mapping F is C^2, $F(\lambda, 0) = 0$ for any λ and

$$F_u(\mu, v)w = -\Delta w + Bw + f'(v)w - \mu w$$

$$\dot{F}_{\lambda u}(\mu, v)w = -w.$$

Hence we obtain

$$F_u(\hat{\lambda}_1 + f'(0), 0)w = -\Delta w + Bw - \hat{\lambda}_1 w$$

and $\text{Ker}(F_u(\hat{\lambda}_1 + f'(0), 0)) = [\varphi_1]$. We choose as a supplementary subspace

$$Z = R(F_u(\hat{\lambda}_1 + f'(0), 0)) = \{z \in C^\alpha(\overline{\Omega}) | \int_\Omega z\varphi_1 = 0\}.$$

It follows that $F_{\lambda u}(\hat{\lambda}_1 + f'(0), 0)\varphi_1 = -\varphi_1 \notin Z$ and by the local inversion theorem (Crandall and Rabinowitz 1973) there exist an interval J containing 0 and C^1 maps $\lambda : J \to \mathbf{R}$ and $\psi : J \to Z$ such that $\psi(0) = 0$, $\lambda(0) = \hat{\lambda}_1 + f'(0)$, $u(s) = s\varphi_1 + s\psi(s)$ implies $F(\lambda(s), u(s)) = 0$ and every nontrivial solution in a neighbourhood of $\hat{\lambda}_1 + f'(0)$ is of this form. Reasoning as in (Hernández 1987, 1988), it can be proved that these solutions are positive (resp. negative) for $s > 0$ (resp. $s < 0$) small and that they can be parameterized by λ in a right neighbourhood of $\hat{\lambda}_1 + f'(0)$. Moreover, for $s > 0$ small, $u(s)$ belongs to the open cone

$$K = \{u \in C_0^1(\overline{\Omega}) | u > 0 \text{ in } \Omega, \frac{\partial u}{\partial n} < 0 \text{ on } \partial\Omega\}.$$

Remark 3.1 The proof of the preceding result involves the uniqueness in Theorem 2.1. It is possible to give an alternative uniqueness proof more in the spirit of the methods in this section, see the proof of Theorem 3.2 below.

We would like to extend to the right the branch of positive solutions given by the local inversion theorem. For this, we need the following auxiliary result.

Lemma 3.1 *Suppose that the assumptions in Theorem 2.1 and*

(3.1) $$f'(\infty) - f'(0) \le \hat{\lambda}_2 - \hat{\lambda}_1$$

are satisfied. If $v > 0$ is a solution of (2.5), then v is nondegenerate, i.e., the derivative $F_u(\lambda, v)$ is an isomorphism.

Proof Let $F(\lambda, v) = 0$. We have

(3.2)
$$-\Delta v + Bv + \frac{f(v)}{v} v = \lambda v \quad \text{in } \Omega$$
$$v = 0 \quad \text{on } \partial\Omega.$$

We claim that $F_u(\lambda, v)$ is an isomorphism. If not, it follows from well-known results that there exists $w \not\equiv 0$ such that

(3.3)
$$-\Delta w + Bw + f'(v)w = \lambda w \quad \text{in } \Omega$$
$$w = 0 \quad \text{on } \partial\Omega.$$

A comparison argument involving Lemma 1.1, (2.3), (3.2) and (3.3) yields

$$\lambda \ge \hat{\lambda}_1(f'(v)) > \hat{\lambda}_1\left(\frac{f(v)}{v}\right)$$

and this is impossible if $\lambda = \hat{\lambda}_1\left(\frac{f(v)}{v}\right)$. A similar comparison argument provides the contradiction.

Remark 3.2 For $B \equiv 0$ the fact that $\lambda = \hat{\lambda}_1\left(\frac{f(v)}{v}\right)$ follows directly from (2.3) because it is the only eigenvalue having a positive eigenfunction (Rabinowitz 1971 a; Hernández 1986). But, to the best of our knowledge, this result is not available for the operator $-\Delta + B + a(x)I$ with nonconstant $a(x)$. This result would follow from the Maximum Principle by Krein-Rutman's Theorem but (see Remark 1.3) we find the same difficulty there. Assumptions like (3.1) are "natural" for concave f but not for convex f (Amann 1976 a, b; Amann and Laetsch 1976; Hernández 1986).

Coming back to the existence proof, the small branch of positive solutions emanating from $\hat{\lambda}_1 + f'(0)$ and going to the right can be extended by using Lemma 3.1 and the Implicit Function Theorem. Using Propositions 18.1 and 19.1 in (Amann 1976 a) the following result can be proved.

Lemma 3.2 *Suppose that the assumptions in Theorem 2.1 are satisfied. Then $\hat{\lambda}_1 + f'(0)$ (resp. $\hat{\lambda}_1 + f'(\infty)$) is the only bifurcation point (resp. the only asymptotic bifurcation point) for positive solutions to (2.5).*

This, together with a continuation argument, may be used to prove the following theorem.

Theorem 3.1 *Suppose that the assumptions of Theorem 2.1 and (3.1) are satisfied. Then for any λ such that $\hat{\lambda}_1 + f'(0) < \lambda < \hat{\lambda}_1 + f'(\infty)$ there exists a unique nondegenerate nontrivial positive solution $u(\lambda)$ for (2.5). Moreover, the mapping $\lambda \to u(\lambda)$ from $(\hat{\lambda}_1 + f'(0), \hat{\lambda}_1 + f'(\infty))$ into $C^\alpha(\overline{\Omega})$ is C^2 and*

$$\lim_{\lambda \uparrow \hat{\lambda}_1 + f'(\infty)} \|u(\lambda)\|_\alpha = +\infty.$$

This local inversion and continuation method has the advantage that it can be applied to the problem (2.5) if the convexity assumption (2.3) is replaced by the concavity assumption

(3.4) $\quad$ the function $\dfrac{f(u)}{u}$ is strictly decreasing for $u > 0$ and strictly increasing for $u < 0$.

It is well-known that in the case $B \equiv 0$ the method of sub and supersolutions is not applicable (as solutions are unstable) and should be replaced by continuation or global bifurcation arguments (Amann 1976 a, b).

The second part of this section is devoted to the study of equation (2.5) under assumptions (2.2), (2.4) and (3.4). It is clear that $0 \leq f'(\infty) < f'(0) < +\infty$. Reasoning as in Lemmas 2.1 and 2.2 we prove the auxiliary results.

Lemma 3.3 *Assume that f satisfies (2.2), (2.4) and (3.4). If $u \in C(\overline{\Omega})$, $u \geq 0$ is a nontrivial solution of (2.5), then $\hat{\lambda}_1 + f'(\infty) < \lambda < \hat{\lambda}_1 + f'(0)$.*

Lemma 3.4 *Assume that f satisfies (2.2), (2.4),(3.4), (1.6) and*

(3.5) $\quad f'(0) - f'(\infty) \leq \hat{\lambda}_1 + \gamma - 2\sqrt{\delta}.$

If $u \in C(\overline{\Omega})$, $u \geq 0$ is a nontrivial solution for (2.5), then $u > 0$ and $\dfrac{\partial u}{\partial n} < 0$ on $\partial \Omega$.

The proof is very similar to the one for Theorem 3.1, but there is a difficulty with uniqueness of positive solutions. More precisely, the local inversion theorem can be applied once again to get a small curve of nontrivial positive solutions bifurcating from $\hat{\lambda}_1 + f'(0)$, this time to the left. At this point, uniqueness is necessary to show that this curve can be parametrized by λ. But the proof of Theorem 2.1 involves sub and supersolutions, which are not available here. An alternative proof is given by exploiting the "uniqueness" implied by the local inversion theorem.

Lemma 3.5 *Suppose that f satisfies the assumptions in Lemma 3.4 and*

(3.6) $\quad f'(0) - f'(\infty) \leq \hat{\lambda}_2 - \hat{\lambda}_1.$

If $v > 0$ is a solution for (2.5), then v is nondegenerate.

Lemma 3.6 *Suppose that the assumptions in Lemma 3.5 are satisfied and that $v > 0$ is a solution of (2.5) for $\mu \in (\hat{\lambda}_1 + f'(\infty), \hat{\lambda}_1 + f'(0))$. Then there exists a smooth curve $u(\lambda)$ of nontrivial positive solutions containing (μ, v) which is defined for $\mu \leq \lambda < \hat{\lambda}_1 + f'(0)$ and such that $u(\lambda) \to 0$ as $\lambda \uparrow \hat{\lambda}_1 + f'(0)$.*

Theorem 3.2 *Suppose that f satisfies the assumptions of Lemma 3.5. Then there exists at most one nontrivial positive solution for (2.5).*

Proof Assume that $u, v \geq 0$ are solutions of (2.5) for some μ. By Lemma 3.4, $u, v > 0$. By Lemma 3.6 there exist two smooth curves $u(\lambda)$ and $v(\lambda)$ passing through (μ, u) and (μ, v) and going to $\hat{\lambda}_1 + f'(0)$. It follows from Lemma 3.5 that these curves cannot intersect and so there must exist two nontrivial positive solutions in a small neighbourhood of $\hat{\lambda}_1 + f'(0)$ which is impossible because of the local inversion theorem.

Theorem 3.3 *Suppose that f satisfies the assumptions of Lemma 3.5. Then, for any λ such that $\hat{\lambda}_1 + f'(\infty) < \lambda < \hat{\lambda}_1 + f'(0)$ there exists a unique nontrivial positive solution $u(\lambda)$ for (2.5). Moreover, the mapping $\lambda \to u(\lambda)$ from $(\hat{\lambda}_1 + f'(\infty), \hat{\lambda}_1 + f'(0))$ into $C^\alpha(\overline{\Omega})$ is C^2 and*

$$\lim_{\lambda \downarrow \hat{\lambda}_1 + f'(\infty)} \|u(\lambda)\|_\alpha = +\infty.$$

Remark 3.3 The case of f satisfying all the assumptions of Theorem 3.3 except (3.6) is partially open even for $B \equiv 0$ (Hernández 1977, 1986). Existence can be proved by using the results in (Amann 1973), but uniqueness is only proved for some interval of λ's.

Remark 3.4 The methods in this section can be applied in other situations. For example, it can be shown that solutions for the problem in Remark 2.5 form a smooth curve (Hernández 1989).

4 The sublinear case: sub and supersolutions, continuation, global bifurcation

We consider the general case of equation (1.2) with sublinear g. First, we give some existence theorems for general g and then specialize to $g(u) = \lambda u - u^3$. See (de Figueiredo and Mitidieri 1986) for these results and related questions.

Assume that g is a C^1 map satisfying

(4.1) $$\lim_{|u| \to +\infty} \frac{g(u)}{u} < \hat{\lambda}_1$$

and

(4.2) $$\lim_{u \to 0} \frac{g(u)}{u} > \hat{\lambda}_1.$$

Remark 4.1 Condition (4.2) is satisfied if, for example, $g(0) > 0$ or $g(0) = 0$ and $g'(0) > \hat{\lambda}_1$.

Theorem 4.1 *Suppose that g is C^1 and that assumptions (1.6), (4.1) and (4.2) are satisfied. If moreover*

(4.3) $$\inf_{u \in [0,b)} g'(u) \geq -\gamma + 2\sqrt{\delta},$$

where $b \leq +\infty$ is the first positive zero of g, then there exists a nontrivial positive solution for (1.2).

Proof It is easy to see, by using (4.2), that $u_0 \equiv c\varphi_1$, for $c > 0$ small enough, is a subsolution. If $b < +\infty$, $u^0 \equiv b > 0$ is a supersolution. If $b = +\infty$, it follows from (4.1) that there exist constants a and $C > 0$ such that $-\gamma + 2\sqrt{\delta} < a < \hat{\lambda}_1$ and $g(u) \leq au + C$; a supersolution is given by the solution of the linear equation $-\Delta z + Bz = az + C$ in Ω, $z = 0$ on $\partial\Omega$.

The method of sub and supersolutions can be used provided there is a $\bar{\mu}$ such that $-\Delta + B - \bar{\mu}I$ satisfies the Maximum Principle and $g(u) - \bar{\mu}u$ is increasing for $0 \leq u \leq \max u^0$. Hence, by Theorem 1.1 and (4.3), $\bar{\mu} = -\gamma + 2\sqrt{\delta}$ is a suitable choice.

Now we consider the particular case $g(u) = \lambda u - u^3$. It is obvious that (4.1) and (4.2) hold for $\lambda > \hat{\lambda}_1$, that $b = \sqrt{\lambda}$ and $\inf_{[0,\sqrt{\lambda})} g'(u) = -2\lambda$. By Theorem 4.1 the equation

(4.4) $$\begin{aligned} -\Delta u + Bu + u^3 &= \lambda u \quad \text{in } \Omega \\ u &= 0 \quad \text{on } \partial\Omega \end{aligned}$$

possesses at least one nontrivial positive solution $u(\lambda)$ for any λ in the interval

(4.5) $$\hat{\lambda}_1 < \lambda < \frac{\gamma}{2} - \sqrt{\delta}.$$

Moreover, uniqueness can be proved as in Theorem 2.1.

Thus equation (4.4) has a unique nontrivial positive solution for any λ in the interval $(\hat{\lambda}_1, \lambda^*)$, where $\lambda^* = \frac{\gamma}{2} - \sqrt{\delta}$. The question arises as to whether or not the critical value λ^* given above is optimal. It is well-known that for $B \equiv 0$ there is a unique nontrivial positive solution for any $\lambda > \lambda_1$ (Sattinger 1972; Hernández 1977, 1986; Berestycki 1981). We remark that sub and supersolutions are still available for $\lambda > \hat{\lambda}_1$ and thus the only difficulty is the application of the Maximum Principle.

The local inversion and continuation in section 3 may also be applied to (4.4). As above, a small curve of nontrivial positive solutions bifurcating to the right from $\hat{\lambda}_1$ is obtained. Comparison arguments as in Lemmas 3.1 and 3.5 show that positive solutions are nondegenerate for $\lambda \in (\hat{\lambda}_1, \hat{\lambda}_2)$, but we do not know how to deal with $\lambda > \hat{\lambda}_2$, see Remarks 3.2 and 1.3. Note that if the result mentioned in Remark 3.2 is true, then existence, uniqueness and nondegeneracy of nontrivial positive solutions would follow, as for $B \equiv 0$.

Remark 4.2 The preceding observations again show the decisive role played by the Maximum Principle and related results. For the problem in (4.4), the two methods give different intervals for existence and uniqueness of positive solutions, but both depend heavily on the Maximum Principle.

Remark 4.3 Lazer and McKenna (1982) have shown that there are exactly three solutions of (4.4) if $\lambda \in (\hat{\lambda}_1, \hat{\lambda}_2)$, but do not discuss the positivity of solutions. They use degree theory in their proof. It is now clear that, besides the trivial solution, there is a nontrivial positive (resp. negative) solution. See also Lemma 2 in (Lazer and McKenna 1982) concerning nondegeneracy of solutions.

Remark 4.4 The function u^3 has been chosen for the sake of simplicity, but analogous results could be obtained for not necessarily odd f satisfying (2.2), (2.3) and

$$(4.6) \qquad \lim_{|u| \to +\infty} \frac{f(u)}{u} = +\infty.$$

The existence of nontrivial positive solutions for (4.4) for any $\lambda > \hat{\lambda}_1$ can be proved by global bifurcation arguments together with *a priori* estimates. Indeed, it is not difficult to see that if $u \geq 0$ is a solution for the value λ of the parameter, then $0 \leq u \leq \sqrt{\lambda}$ (Rabinowitz 1971 a; Hernández 1988). On the other hand, if $\hat{\lambda}_1 < \frac{\gamma}{2} - \sqrt{\delta}$, then $\gamma > 2\sqrt{\delta}$ and the Maximum Principle holds for $\mu = 0$ by Theorem 1.1. Thus equation (4.4) is equivalent to $u = T_0(\lambda u - u^3)$ and by the above considerations positive solutions for this problem are the same as for $u = T_0(g(\lambda, u))$, where

$$g(\lambda, u) = \begin{cases} \lambda u - u^3 & \text{if } |u| \leq \sqrt{\lambda} \\ 0 & \text{if } |u| \geq \sqrt{\lambda}. \end{cases}$$

It is easily seen that $\tilde{f}(\lambda, u) = T_0(g(\lambda, u))$ satisfies the assumptions of Corollary 18.4 in (Amann 1976 a) and this provides the existence of an unbounded continuum of positive solutions bifurcating from $\hat{\lambda}_1$. This together with the *a priori* estimates for positive solutions for any λ proves existence. On the other hand, uniqueness remains unsettled for $\lambda > \max\{\lambda^*, \hat{\lambda}_2\}$.

Remark 4.5 J. Blat (personal communication) has obtained an alternative proof by using sub and supersolutions as in, e.g., (Hernández 1986; Section 1.3).

Remark 4.6 Other existence and multiplicity results for positive solutions have been proved by Sobreira(1985 a) again using global bifurcation (Rabinowitz 1971 b, c). Roughly speaking, his results concern the case of g asymptotically linear with $g(0) \geq 0$. For the case of Neumann boundary conditions, see (Sobreira 1985 b).

Remark 4.7 Results for the superlinear case have been obtain by Brezis and Turner (1977) and de Figueiredo et al. (1982) for a single equation. De Figueiredo and Mitidieri (1986) obtained nonexistence results by extending the Pohozaev identity to systems and Sobreira (1985 a) gives existence and multiplicity results by exploiting in a nontrivial way global bifurcation. See Sobreira (1985 b) for the Neumann problem.

5 The sublinear case: critical point theory

We now study equation (1.2) again under a different set of assumptions and exploiting another technique, namely critical point theory. Most of the results follow as applications of the well-known Mountain Pass Theorem (Ambrosetti and Rabinowitz, 1973).

The nonlinearity g is C^1 and satisfies (4.1) together with either

$$\text{(5.1)} \qquad \lim_{|u|\to+\infty} \frac{g(u)}{|u|^p} = 0$$

with $1 < p < \frac{N+2}{N-2}$ for $N \geq 3$, $1 < p < \infty$ if $N = 2$, or

$$\text{(5.2)} \qquad \lim_{|u|\to+\infty} \frac{g(u)}{u} < -\frac{\delta}{\gamma}.$$

The study of solutions of (1.2) is equivalent to finding the critical points of the functional

$$\Phi(u) = \frac{1}{2}\int_\Omega |\nabla u|^2 + u \cdot Bu - G(u)$$

where $G(u) = \int_0^u g(s)ds$. The functional Φ is well-defined in $H_0^1(\Omega)$ if g satisfies (5.1), but not if g satisfies (5.2) where some truncation of g is necessary. A modified Φ which is well-defined in $H_0^1(\Omega)$ and bounded below is obtained by exploiting a *priori* estimates for solutions of (1.2). This functional is C^1 and gives solutions in $H_0^1(\Omega)$; these solutions can be proved to be classical solutions by standard bootstrapping arguments.

First, we obtain the *a priori* estimates. Cases (5.1) and (5.2) are treated separately. The first is considered in (de Figueiredo and Mitidieri 1986) and the second in (Rothe 1981; Lazer and McKenna 1982; de Figueiredo and Mitidieri 1986).

Lemma 5.1 *If g satisfies* (4.1) *and* (5.1), *then there exists $C > 0$ such that for any solution of* (1.2), $\|u\|_\infty \leq C$.

Proof It follows from (4.1) that there exist constants a, M with $0 < a < \hat{\lambda}_1$ and $M > 0$ such that

$$\text{(5.3)} \qquad g(s) \leq as + M \quad \text{if } s \geq 0,$$

$$g(s) \geq as - M \quad \text{if } s \leq 0.$$

Multiplying (1.2) by u, integrating over Ω by using Green's Formula and inequality (1.8) in (de Figueiredo and Mitidieri 1986), we obtain

(5.4) $$\lambda_1 \int_\Omega u^2 \leq \int_\Omega |\nabla u|^2 + u \cdot Bu = \int_\Omega g(u)u$$

and by (5.3)

(5.5) $$\int_\Omega g(u)u \leq a \int_\Omega u^2 + M \int_\Omega |u|,$$

which implies $\int_\Omega u^2 \leq C$. (We use C to denote various different constants). It now follows from (5.4) and (5.5) that $\int_\Omega |\nabla u|^2 \leq C$. By (5.1) for every $\varepsilon > 0$ there is a $C_\varepsilon > 0$ such that

$$|g(s)| \leq \varepsilon |s|^p + C_\varepsilon$$

and so the result follows from the Sobolev Imbedding Theorem and L^p regularity results.

Lemma 5.2 *If g satisfies (4.1) and (5.2), then there exists $C > 0$ such that for any solution of (1.2), $\|u\|_\infty \leq C$.*

Proof We prove first the inequality

(5.6) $$\frac{\delta}{\gamma} \min u \leq Bu(x) \leq \frac{\delta}{\gamma} \max u$$

for every $u \in C(\overline{\Omega})$ with $u = 0$ on $\partial\Omega$. By definition $v = Bu$ satisfies

(5.7) $$v = \frac{\delta}{\gamma} u + \frac{1}{\gamma} \Delta v.$$

We only prove the first part of (5.6), the second is very similar. The result is obvious for $v \geq 0$. If not, we have $\min v = v(x_1) < 0$ for some $x_1 \in \Omega$. Then $\Delta v(x_1) \geq 0$ and (5.7) implies $v(x_1) \geq \frac{\delta}{\gamma} u(x_1)$, which immediately gives the result.

By (5.2) there exist constants a and m such that

(5.8) $$\frac{g(s)}{s} \leq -a < -\frac{\delta}{\gamma} \text{ if } |s| \geq m.$$

We claim that $\|u\|_\infty \leq m$ for every solution u. Suppose we have $\|u\|_\infty = u(x_0) = M > m$ for some $x_0 \in \Omega$. Then by (5.6), (5.8) and (4.1)

$$-\frac{\delta}{\gamma} M \leq g(u(x_0)) \leq -au(x_0) = -aM,$$

a contradiction. The proof for $u(x_1) = -M$ is analogous.

Theorem 5.1 (*de Figueiredo and Mitidieri 1986*). *Assume that g satisfies (4.1) and (5.1) (or (5.2)). Then there exists at least one solution for (1.2).*

Proof (Sketch) Φ is a C^1 functional which is bounded below and satisfies the Palais-Smale condition. Hence Φ has a global minimum.

Remark 5.1 The result becomes trivial if $g(0) = 0$. On the other hand, the example $g(u) = au$ with $a < \lambda_1$ shows that the trivial solution may be unique.

The following multiplicity theorem can be proved by using the Mountain Pass Theorem.

Theorem 5.2 *Assume that g satisfies* (4.1) , (5.1) *(or* (5.2)*) and*

(5.9) $$g(0) = 0, \; g'(0) < \hat{\lambda}_1.$$

Moreover, assume that there is an $a > 0$ such that $G(a) > G(s)$ for $s \in [0, a]$ and that

(5.10) $$\frac{2G(a)}{a^2} > \min\{\frac{1}{R^2}\frac{(1+t)^2}{t^2}\frac{(1+t)^N - 1}{2 - (1+t)^N}$$
$$+ \frac{\delta}{\gamma}\frac{(1+t)^N}{2 - (1+t)^N} : 0 < t < 2^{\frac{1}{N}} - 1\},$$

where R is the radius of the largest ball contained in Ω. Then there are at least two nontrivial solutions of (1.2).

Remark 5.2 Condition (5.10) is satisfied if $G(b) > 0$ for some $b > 0$ and , for example, Ω is very large ball and δ is very small. The case $g(u) = u(u - a)(1 - u)$, $0 < a < 1$, was studied first, by using essentially the same methods (Klaasen and Mitidieri 1986; Klaasen 1987). In particular, the first paper gives conditions less general than (5.10) ensuring the existence of two nontrivial solutions, and some nonexistence results are also given.

A different, and more difficult, problem which can also be treated by using critical points arguments is the system

$$-\Delta u = \lambda u - \delta v + f(u) \quad \text{in } \Omega$$

$$-\Delta v = \delta u - \gamma v - h(v) \quad \text{in } \Omega$$

$$u = v = 0 \quad \text{on } \partial\Omega$$

where f and h are such that the system is sublinear in v but superlinear in u. The case $h(v) = v^3$ and $f(u) = |u|^{N+2/N-2} \operatorname{sgn} u$ has been considered by Mancini and Mitidieri (1987). They obtain existence and nonexistence results which extend in some directions the work of Brezis and Nirenberg (1983). We recall that $\frac{N+2}{N-2}$ is the critical Sobolev exponent and thus this is a critical problem for systems. Again, the Maximum Principle plays a leading role in the proofs. See also (Caristi and Mitidieri 1988) for nonexistence results. Symmetry properties are also studied in (Mancini and Mitidieri 1987).

Remark 5.3 Some multiplicity theorems which hold for the case of a single equation may be extended to certain systems, see, e.g., (Ambrosetti and Mancini 1979; Struwe 1982).

6 Nonlinear decoupling: global bifurcation

In this last section we consider a more difficult problem with two nonlinear equations where decoupling becomes less obvious. We study the system

(6.1)
$$-\Delta u = au - f(u) - uv \quad \text{in } \Omega$$
$$-\Delta v = bv - g(v) + uv \quad \text{in } \Omega$$
$$u = v = 0 \quad \text{on } \partial\Omega,$$

where f and g satisfy (2.2), (2.3) and (4.6), a and b are real parameters. The system (6.1) describes a predator-prey interaction, where u denotes the prey population and v the predator population (Conway et al., 1982; Blat and Brown 1984).

First, we fix a and take b as a bifurcation parameter. In order to describe the system we consider the equation satisfied by u for fixed v in $C^1(\overline{\Omega})$. We have

(6.2)
$$-\Delta u + f(u) + vu = au \quad \text{in } \Omega$$
$$u = 0 \quad \text{on } \partial\Omega.$$

The equation (6.2) has the trivial solution $u \equiv 0$ for every a and this solution is unique for $a \leq \lambda_1(v)$; for $a > \lambda_1(v)$ there is also a unique nontrivial positive solution $u(a) > 0$ (Sattinger 1972; Hernández 1986).

We define a mapping $B : C^1(\overline{\Omega}) \to C^1(\overline{\Omega})$ in the following way:

$$B(v) = \begin{cases} 0 & \text{if } a \leq \lambda_1(v), \\ u(a) & \text{if } a > \lambda_1(v). \end{cases}$$

In this way the study of positive solutions for (6.1) is equivalent to the study of positive solutions for the nonlinear equation

(6.3)
$$-\Delta v + g(v) - vB(v) = bv \quad \text{in } \Omega$$
$$v = 0 \quad \text{on } \partial\Omega.$$

As in the preceding sections, results for (6.3) are obtained and then "translated" to (6.1). First, we remark that a simple argument involving the Maximum Principle shows that if $v \geq 0$ is a solution for (6.3), then $v > 0$.

It is obvious that B is a nonlinear mapping and this raises an additional difficulty with respect to the problems considered in the previous sections, where B was a nice compact self-adjoint linear operator. However, B possesses some properties which make it possible to apply general global bifurcation results (Blat and Brown 1984).

Lemma 6.1 $B: C^1(\overline{\Omega}) \to C^1(\overline{\Omega})$ *is continuous and order-reversing ($u \leq v$ implies $Bu \geq Bv$).*

Consider first the trivial cases where $u \equiv 0$ or $v \equiv 0$. For $u \equiv 0$, v is a solution of the nonlinear equation

(6.4)
$$-\Delta v + g(v) = bv \quad \text{in } \Omega$$
$$v = 0 \quad \text{on } \partial\Omega,$$

which has the trivial solution $v \equiv 0$ and a unique nontrivial positive solution $v(b) > 0$ for $b > \lambda_1$. Analogously, for $v \equiv 0$ we have necessarily $u \equiv 0$ for $a \leq \lambda_1$, and $u \equiv 0$ and $u(a) > 0$ for $a > \lambda_1$.

Thus all solutions are known when $a \leq \lambda_1$. We now consider the case $a > \lambda_1$ fixed. We point out that in this case $B(0) = u(a) > 0$. For any b we have the trivial solution $(0,0)$ and the semi-trivial solution $(B(0), 0)$ to (6.1). Now the main goal is the study of bifurcation of positive solutions of (6.1) with respect to the semi-trivial solution $(B(0), 0)$ or, equivalently, bifurcation of positive solutions for (6.3) with respect to the trivial solution $v \equiv 0$. The corresponding linearized problem is

$$-\Delta w - B(0)w = bw \quad \text{in } \Omega$$
$$w = 0 \quad \text{on } \partial\Omega,$$

and there is no loss of generality in assuming that the operator $-\Delta w - B(0)w$ is invertible, i.e., that 0 is not an eigenvalue.

The associated Nemitskii operator $F: C^1(\overline{\Omega}) \to C^1(\overline{\Omega})$ is defined by

$$F(v) = -g(v) + (B(v) - B(0))v.$$

It follows from Lemma 6.1 that F is continuous and $F(v) = o(\|v\|)$ at $v = 0$.

If $K = (-\Delta - B(0)I)^{-1}$ then K is a compact linear operator in $C^1(\overline{\Omega})$ and (6.3) may be rewritten as

(6.5)
$$v = bKv + KF(v).$$

Since $F(v) = o(\|v\|)$ and $b = \lambda_1(-B(0))$ is a simple characteristic value of K, well-known global bifurcation results (Rabinowitz 1971 b, c; Amann 1976 a) may be applied to obtain the existence of a continuum C^+ of positive solutions for (6.5) bifurcating at $\lambda_1(-B(0))$ and satisfying one of the alternatives in Rabinowitz's Theorem. Recall that

$$P = \{u \in C_0^1(\overline{\Omega}): u > 0 \text{ in } \Omega, \frac{\partial u}{\partial n} < 0 \text{ on } \partial\Omega\}$$

is an open cone.

Theorem 6.1 *We have $C^+ \subset \mathbf{R}^+ \times P$. Moreover, the projection of C^+ on $\mathbf{R}^+$ is the interval $(\lambda_1(-B(0)), +\infty)$.*

Proof (Sketch) The proof of the first part is similar to arguments used in the proof of Theorem 3.1. An analogous argument implies that C^+ cannot be bounded. If $(\mu, v) \in C^+$, with $v > 0$, a comparison argument involving Lemmas 6.1 and 1.1 gives $\mu \geq \lambda_1(-B(0))$. Now to complete the proof it is sufficient to find a priori estimates for positive solutions. Let $M(b)$ be a constant such that $bv - g(v) + B(0)v < 0$ for $v > M(b)$ and let $G = \{x \in \Omega | v(x) > M(b)\}$. By Lemma 6.1, $-\Delta v = bv - g(v) + B(v)v \leq 0$ and $v = M(b)$ on ∂G, and it follows from the Maximum Principle that $v(x) \leq M(b)$, a contradiction. By well-known classical regularity results there exists a constant $C(b)$ such that $\|v\|_{C^1} \leq C(b)$.

Solutions $v > 0$ are "small" near the bifurcation point $\lambda_1(-B(0))$, and the continuity of B implies $u = B(v) > 0$. But as b increases, it could happen that v increases, until $B(v) \equiv 0$; in other words, v may increase until a solution (u, v) of (6.1) is reached with $u \equiv 0$. The following result shows that this must occur.

Theorem 6.2 *For any $a > \lambda_1$ fixed, there is a constant $C(a) > 0$ such that for $b > C(a)$ there is no solution of (6.1) with $u > 0$ and $v > 0$.*

Proof Let (u, v) be a solution for (6.1). If $v > 0$, then it is the unique nontrivial positive solution of

(6.6)
$$-\Delta v + g(v) - uv = bv \quad \text{in } \Omega$$
$$v = 0 \quad \text{on } \partial\Omega.$$

It is not difficulty to see that for $b > \lambda_1$, $k(b)\varphi_1$, where $k(b) = b - \lambda_1$, is a subsolution for (6.6) and that $C > 0$ large enough is a supersolution. Hence $v \geq k(b)\varphi_1$ with $k(b) \to +\infty$ as $b \to +\infty$.

Consider the eigenvalue problem

$$-\Delta w + k(b)\varphi_1 w = \mu w \quad \text{in } \Omega$$
$$w = 0 \quad \text{on } \partial\Omega.$$

We claim that $\lambda_1(k(b)\varphi_1) \to +\infty$ as $b \to +\infty$. If not, it follows from the variational characterization of the eigenvalues that there exists a sequence u_n of functions in $H_0^1(\Omega)$ with $\|u_n\|_{L^2} = 1$, $\|\nabla u_n\|_{L^2}$ is uniformly bounded and $\int_\Omega \varphi_1 u_n^2 \to 0$ as $n \to +\infty$. Then there exists a convergent subsequence $u_n \to u_0$ in $L^2(\Omega)$, and hence $\int_\Omega \varphi_1 u_0^2 = 0$. Thus $u_0 \equiv 0$, a contradiction to $\|u_0\|_{L^2} = 1$.

We may choose b large enough so that $\lambda_1(k(b)\varphi_1) > a$. Another comparison argument gives $\lambda_1(v) > \lambda_1(k(b)\varphi_1) > a$, and this implies $u = Bv \equiv 0$. This completes the proof.

Let us "translate" this branch of solutions from the "plane" (b, v) into the "space" (b, u, v) by means of the mapping $(b, v) \to (b, B(v), v)$. It follows from Theorem 6.2 that if the continuum C^+ "goes to infinity to the right", then $u \equiv 0$ for b large enough. But if $u \equiv 0$, v is a solution for (6.4) and the continuum C^+ provides a solution set for (6.1) containing, for b sufficiently large, only elements of the form $u \equiv 0$ and $v = v(b) > 0$.

Theorem 6.3 *There exists a $\lambda^* > \lambda_1$ such that for any b satisfying $\lambda_1(-B(0)) < b < \lambda^*$, the system (6.1) has at least one solution (u, v) with $u > 0$ and $v > 0$. Moreover, there is also a $\overline{\lambda}$, $\overline{\lambda} \geq \lambda^*$, such that if $b > \overline{\lambda}$, then for every solution for (6.1) $u \equiv 0$ or $v \equiv 0$.*

These results have a reasonable biological meaning. If the birth rate of the predator is too small $(b < \lambda_1(-B(0)))$, predators cannot survive. On the contrary, if it is too large $(b > \overline{\lambda})$, then the prey cannot survive. The meaning of Theorem 6.3 is that if there is some balance between both species, then they may coexist, i.e., there are solutions with $u > 0$ and $v > 0$.

Remark 6.1 Another proof of existence of solutions to (6.1) or, more precisely, to the very closely related system studied in (Conway *et al.* 1982) has been given by Dancer (1984) by using degree for cones. He provides necessary and sufficient conditions, expressed in terms of the spectral radii of some linear operators, for the existence of coexistence solutions. These techniques are much more complicated but, on the other hand, they apply to more general situations (other nonlinearities, different differential operators (see Remark 1.4), systems with more than two equations, etc). Moreover, these techniques can also be used to get multiplicity and stability results (Dancer 1984, 1985).

Remark 6.2 Uniqueness for nontrivial positive solutions seems to be a rather difficult question. To the best of our knowledge, the only available result is by Dancer (1985) for the one-dimensional case. Uniqueness has been proved for the competition system (Cosner and Lazer 1984; Cantrell and Cosner 1987).

A similar study may be carried out by fixing $b > 0$ and taking a as a bifurcation parameter. The decoupling is now achieved by solving for fixed u the equation

(6.7)
$$-\Delta v + g(v) - uv = bv \quad \text{in } \Omega$$
$$v = 0 \quad \text{on } \partial\Omega,$$

which has only the trivial solution $v \equiv 0$ for $b \leq \lambda_1(-u)$ and a unique nontrivial positive solution $v(b) > 0$ if $b > \lambda_1(-u)$. For fixed a, the mapping defined by

$$C(u) = \begin{cases} 0 & \text{if } b \leq \lambda_1(-u) \\ v(b) & \text{if } b > \lambda_1(-u) \end{cases}$$

is continuous and order-preserving ($u \leq v$ implies $C(u) \leq C(v)$) in $C^1(\overline{\Omega})$. The study of solutions with either $u \equiv 0$ or $v \equiv 0$ is carried out as above. A rather similar argument is used to prove the corresponding existence result.

Theorem 6.4 *If $b > \lambda_1$ and $a > \lambda_1(C(0))$, then system (6.1) has solutions with $u > 0$ and $v > 0$.*

When $b \leq \lambda_1$, existence cannot be proved by applying global bifurcation arguments directly, but it is possible to use the previous existence theorems to obtain the following result.

Theorem 6.5 *If $b \leq \lambda_1$, then for $a > 0$ sufficiently large there are solutions to (6.1) with $u > 0$ and $v > 0$.*

Remark 6.3 The competition system obtained by replacing uv by $-uv$ in the second equation of (6.1) can be treated by the same methods. It has been proved (Blat and Brown 1984) that there are coexistence solutions for $a > \lambda_1$ and b in some bounded interval. On the other hand when b is sufficiently large, one of the components should be identically zero. See (Dancer 1984, 1985) for interesting observations on this and related problems.

Remark 6.4 An interesting variant of (6.1) is the Holling-Tanner interaction, which is obtained by putting $\frac{uv}{1+mu}$ instead of uv in both equations of (6.1). (Here $m > 0$ is a given constant). The equation arising in the decoupling of the system is

(6.8)
$$-\Delta u + f(u) + \frac{uv}{1+mu} = au \quad \text{in } \Omega$$
$$u = 0 \quad \text{on } \partial\Omega,$$

with $v \geq 0$ fixed. Existence of nontrivial positive solutions of (6.8) can be proved as above, but uniqueness raises delicate problems. (The nonlinearity in the left-hand of (6.8) is the sum of a convex and a concave function, see (Amann 1976 a; Hernández 1986; Lions 1982) for related results). Moreover, it is possible to apply directly global bifurcation arguments (Rabinowitz 1971 b, c) showing the existence of a continuum of solutions with $u > 0$ and $v > 0$ joining two branches of semi-trivial positive solutions. On the other hand, if we take a as a bifurcation parameter, the decoupling technique described above applies (now the nonlinear term on the left-hand side is convex). See (Blat and Brown 1986 a) for details.

References

[1] Amann, H. (1973). Fixed points of asymptotically linear maps in ordered Banach spaces. *Journal of Functional Analysis*, **14**, 162-71.

[2] Amann, H.(1976 a). Fixed point equations and nonlinear eigenvalue problems in ordered Banach spaces. *SIAM Review*, **18**, 620-709.

[3] Amann, H. (1976 b). Nonlinear eigenvalue problems having precisely two solutions. *Mathematische Zeitschrift*, **150**, 27-37.

[4] Amann, H. and Laetsch, T. (1976). Positive solutions of convex nonlinear eigenvalue problems. *Indiana University Mathematical Journal*, **25**, 259-70.

[5] Ambrosetti, A. and Mancini, G. (1979). Sharp nonuniqueness results for some nonlinear problems. *Nonlinear Analysis*, **3**, 635-45.

[6] Ambrosetti, A. and Rabinowitz, P. H. (1973). Dual variational methods in critical point theory and applications. *Journal of Functional Analysis*, **14**, 349-81.

[7] Berestycki, H. (1981). Le nombre de solutions de certains problèmes semilinéaires elliptiques. *Journal of Functional Analysis*, **40**, 1-29.

[8] Blat, J. (1987). Bifurcation to positive solutions in systems of elliptic equations. In *Contributions to nonlinear partial differential equations*, Vol. II (ed. J. I. Diaz and P. L. Lions), pp. 39-46. Longman, Essex.

[9] Blat, J. and Brown, K. J. (1984). Bifurcation of steady-state solutions in predator-prey and competition systems. *Proceedings of the Royal Society of Edinburgh*, **97 A**, 21-34.

[10] Blat, J. and Brown, K. J. (1986 a). Global bifurcation of positive solutions in some systems of elliptic equations. *SIAM Journal in Mathematical Analysis*, **17**, 1339-53.

[11] Blat, J. and Brown, K. J. (1986 b). A reaction-diffusion system modelling the spreading of bacterial infections. *Mathematical Methods in the Applied Sciences*, **8**, 234-46.

[12] Brown, K. J. (1983). Spatially inhomogeneous steady-state solutions for systems of equations describing interacting populations. *Journal of Mathematical Analysis and Applications*, **95**, 251-64.

[13] Brown, K. J. (1987). Nontrivial solutions of predator-prey systems with small diffusion. *Nonlinear Analysis*, **11**, 685-9.

[14] Brezis, H. and Nirenberg, L. (1983). Positive solutions of nonlinear elliptic equations involving critical exponents. *Communications in Pure and Applied Mathematics*, **34**, 437-77.

[15] Brezis, H. and Turner, R. E. L. (1977). On a class of superlinear elliptic problems. *Communications in Partial Differential Equations*, **2**, 601-14.

[16] Cantrell, R. S. and Cosner, C. (1987). On the uniqueness and stability of positive solutions in the Volterra-Lotka competition model with diffusion. To appear.

[17] Cantrell, R. S. and Schmidt, K. (1986). On the eigenvalue problem for coupled elliptic systems. *SIAM Journal in Mathematical Analysis*, **17**, 850-62.

[18] Capasso, V. and Maddalena, L. (1982). Saddle point behaviour for a reaction-diffusion system modelling the spatial spread of a class of bacterial and viral diseases. *Mathematics and Computers in Simulation*, **24**, 540-7.

[19] Caristi, G. and Mitidieri, E. (1988). Maximum principles for a class of non-cooperative elliptic systems. To appear.

[20] Conway, E., Gardner, R. and Smoller, J. (1982). Stability and bifurcation of steady-state solutions for predator-prey equations. *Advances in Applied Mathematics*, **3**, 288-334.

[21] Cosner, C. (1988). Spectrum, generalized spectrum, and eigenfunctions for second order elliptic systems. In *Maximum principles and eigenvalue problems in partial differential equations* (ed. P. W. Schaeffer), pp. 94-110. Longman, Essex.

[22] Cosner, C. and Lazer, A. (1984). Stable coexistence states in the Volterra-Lotka competition model with diffusion. *SIAM Journal in Applied Mathematics*, **44**, 1112-32.

[23] Courant, R. and Hilbert, D. (1953). *Methods of mathematical physics*. Interscience, New York.

[24] Crandall, M. and Rabinowitz, P. H. (1973). Bifurcation, perturbation of simple eigenvalues and linearized stability. *Archive for Rational Mechanics and Analysis*, **52**, 161-80.

[25] Dancer, E. N. (1984). On positive solutions of some pairs of differential equations I. *Transactions of the American Mathematical Society*, **284**, 729-43.

[26] Dancer, E. N. (1985). On positive solutions of some pairs of differential equations II. *Journal of Differential Equations*, **60**, 236-58.

[27] de Figueiredo, D. G., Lions, P. L. and Nussbaum, R. (1982). A priori estimates and existence of positive solutions of semilinear elliptic equations. *Journal de Mathématiques Pures et Appliquées*, **61**, 41-63.

[28] de Figueiredo, D. G. and Mitidieri, E. (1986). A maximum principle for an elliptic system and applications to semilinear problems. *SIAM Journal in Mathematical Analysis*, **17**, 836-49.

[29] de Figueiredo, D. G. and Mitidieri, E. (1988). Maximum principles for linear elliptic systems. To appear.

[30] Diaz, J. I. (1985). *Nonlinear partial differential equations and free boundaries*. Pitman, London.

[31] Hernández, J. (1977). Bifurcación y soluciones positivas para ciertos problemas de tipo unilateral. Thesis. Universidad Autónoma de Madrid.

[32] Hernández, J. (1986). Qualitative methods for nonlinear diffusion equations. In *Nonlinear diffusion problems* (ed. A. Fasano and M. Primicerio), pp. 47-118, Springer Lecture Notes in Mathematics, **1224**. Springer-Verlag, Berlin.

[33] Hernández, J. (1987). Continuation and comparison methods for some nonlinear elliptic systems. In *Contributions to nonlinear partial differential equations*, Vol. II (ed. J. I. Diaz and P. L. Lions), pp. 137-47. Longman, Essex.

[34] Hernández, J. (1988). Continuation and monotone methods for positive solutions of some reaction-diffusion systems. To appear.

[35] Hernández, J. (1989). Branches of positive solutions for a reaction-diffusion system arising in the spread of bacterial infections. To appear.

[36] Hess, P. (1983). On the eigenvalue problem for weakly coupled elliptic systems. *Archive for Rational Mechanics and Analysis*, **81**, 151-9.

[37] Hirsch, F. (1972). Familles résolvantes, générateurs, cogénerateurs, potentiels. *Annales Institut Fourier*, **22**, 89-210.

[38] Klaasen, G. (1987). Stationary spatial patterns for a reaction-diffusion system with an excitable steady state. To appear.

[39] Klaasen, G. and Mitidieri, E. (1986). Standing wave solutions for a system derived from the Fitzhugh-Nagumo equations for nerve conduction. *SIAM Journal in Mathematical Analysis*, **17**, 74-83.

[40] Lazer, A. and McKenna, P. J. (1982). On steady-state solutions of a system of reaction-diffusion equations from biology. *Nonlinear Analysis*, **6**, 523-30.

[41] Lions, P. L. (1982). On the existence of positive solutions of semilinear elliptic equations. *SIAM Review*, **24**, 441-67.

[42] Mancini, G. and Mitidieri, E. (1987). Positive solutions of some coercive-anticoercive elliptic systems. *Annales de la Faculté des Sciences de Toulouse*, **8**, 257-292.

[43] Martin Jr., R. H. (1976). *Nonlinear operators and differential equations in Banach spaces*. Wiley, New York.

[44] Mirenghi, E. (1987). Equilibrium solutions of a semilinear reaction-diffusion system. *Nonlinear Analysis*, **11**, 393-405.

[45] Protter, M. H. and Weinberger, H. F. (1967). *Maximum principles in differential equations*. Prentice-Hall, New Jersey.

[46] Rabinowitz, P. H. (1971 a). A note on a nonlinear eigenvalue problem for a class of differential equations. *Journal of Differential Equations*, 9, 536-48.

[47] Rabinowitz, P. H. (1971 b). Some global results for nonlinear eigenvalue problems. *Journal of Functional Analysis*, 7, 487-513.

[48] Rabinowitz, P. H. (1971 c). A global theorem for nonlinear eigenvalue problems and applications. In *Contributions to nonlinear functional analysis*. (ed. E. H. Zarantonello), pp. 11-36. Academic Press, New York.

[49] Rothe, F. (1981). Global existence of branches of stationary solutions for a system of reaction-diffusion equations from biology. *Nonlinear Analysis*, 5, 487-98.

[50] Rothe, F. and de Mottoni, P. (1979). A simple system of reaction-diffusion equations describing morphogenesis: asymptotic behaviour. *Annali di Matematica Pura ed Applicata*, 122, 141-57.

[51] Sattinger, D. H. (1972). Monotone methods in nonlinear elliptic and parabolic equations. *Indiana University Mathematical Journal*, 21, 979-1000.

[52] Smoller, J. (1983). *Shock waves and reaction-diffusion equations*. Springer, Berlin.

[53] Sobreira, F. (1985 a). On the existence and multiplicity of positive solutions of a semilinear elliptic system. Preprint.

[54] Sobreira, F. (1985 b). On a elliptic system with Neumann boundary conditions. Preprint.

[55] Sobreira, F. (1986). Sobre a não existencia de um Principio de Máximo para un sistema eliptico. To appear.

[56] Struwe, M. (1982). A note on a result of Ambrosetti and Mancini. *Annali di Matematica Pura ed Applicata*, 131, 107-15.

[57] Sweers, G. (1988). A strong maximum principle for a non-cooperative elliptic system. To appear.

[58] Weinberger, H. F. (1988). Some remarks on invariant sets for systems. In *Maximum principles and eigenvalue problems in partial differential equations* (ed. P. W. Schaefer), pp. 189-207. Longman, Essex.